Foundations of Differential Calculus

Springer
New York
Berlin
Heidelberg
Barcelona
Hong Kong
London
Milan
Paris
Singapore
Tokyo

Euler

Foundations of Differential Calculus

Translated by John D. Blanton

Translator
John D. Blanton
Department of Mathematics
and Computer Science
St. John Fisher College
Rochester, NY 14618
USA
blanton@sjfc.edu

Cover illustration: Portrait of Leonhard Euler, Artist Unknown. Stock Montage/Superstock.

Mathematics Subject Classification (2000): 01A50, 26A24, 26B05, 34-03

Library of Congress Cataloging-in-Publication Data
Euler, Leonhard, 1707–1783.
[Institutiones calculi differentialis. Chapter 1–9. English]
Foundations of differential calculus/Euler; translated by John D. Blanton.
p. cm.
Includes bibliographical references and index.
ISBN 0-387-98534-4 (alk. paper)
1. Differential calculus—Early works to 1800. I. Title.
QA302 .E8513 2000
515′.3—dc21 99-043386

Printed on acid-free paper.

Translated from the Latin *Institutiones Calculi Differentialis*, Chapters 1 to 9, by Leonhard Euler, 1755.

Production managed by Francine McNeill; manufacturing supervised by Erica Bresler.
Typeset by David Kramer, Lancaster, PA, from the translator's TeX files.
Printed and bound by Edwards Brothers, Inc., Ann Arbor, MI.
Printed in the United States of America.

9 8 7 6 5 4 3 2 1

ISBN 0-387-98534-4 Springer-Verlag New York Berlin Heidelberg SPIN 10678538

Preface

What differential calculus, and, in general, analysis of the infinite, might be can hardly be explained to those innocent of any knowledge of it. Nor can we here offer a definition at the beginning of this dissertation as is sometimes done in other disciplines. It is not that there is no clear definition of this calculus; rather, the fact is that in order to understand the definition there are concepts that must first be understood. Besides those ideas in common usage, there are also others from finite analysis that are much less common and are usually explained in the course of the development of the differential calculus. For this reason, it is not possible to understand a definition before its principles are sufficiently clearly seen.

In the first place, this calculus is concerned with variable quantities. Although every quantity can naturally be increased or decreased without limit, still, since calculus is directed to a certain purpose, we think of some quantities as being constantly the same magnitude, while others change through all the stages of increasing and decreasing. We note this distinction and call the former *constant quantities* and the latter *variables.* This characteristic difference is not required by the nature of things, but rather because of the special question addressed by the calculus.

In order that this difference between constant quantities and variables might be clearly illustrated, let us consider a shot fired from a cannon with a charge of gunpowder. This example seems to be especially appropriate to clarify this matter. There are many quantities involved here: First, there is the quantity of gunpowder; then, the angle of elevation of the cannon above the horizon; third, the distance traveled by the shot; and, fourth, the length of time the shot is in the air. Unless the same cannon is used throughout the experiment, we must also bring into our calculations the length of the barrel and the weight of the shot. Here, we will not consider variations in the cannon or the shot, lest we become entailed in very complicated questions. Hence, if we always keep the same quantity of powder, the elevation of the barrel will vary continuously with the distance traveled and the shot's duration of time in the air. In this case, the amount of powder, or the force of the explosion, will be the constant quantity. The elevation of the barrel, the distance traveled, and the time in the air should be the variable quantities. If for each degree of elevation we were to define these things, so that they may be noted for future reference, the changes in distance and duration of the flight arise from all of the different elevations. There is another question: Suppose the elevation of the barrel is kept the same, but the quantity of powder is continuously changed. Then the changes that occur in the flight need to be defined. In this case, the elevation will be the constant, while the quantity of powder, the distance, and duration are the variable quantities. Hence, it is clear that when the question is changed, the quantities that are constant and those that are variables need to be noted. At the same time, it must be understood from this that in this business the thing that requires the most attention is how the variable quantities depend on each other. When one variable changes, the others necessarily are changed. For example, in the former case considered, the quantity of powder remains the same, and the elevation is changed; then the distance and duration of the flight are changed. Hence, the distance and duration are variables that depend on the elevation; if this changes, then the others also change at the same time. In the latter case, the distance and duration depend on the quantity of charge of powder, so that a change in the charge must result in certain changes in the other variables.

Those quantities that depend on others in this way, namely, those that undergo a change when others change, are called *functions* of these quantities. This definition applies rather widely and includes all ways in which one quantity can be determined by others. Hence, if x designates the variable quantity, all other quantities that in any way depend on x or are determined by it are called its functions. Examples are x^2, the square of x, or any other powers of x, and indeed, even quantities that are composed with these powers in any way, even transcendentals, in general, whatever depends on x in such a way that when x increases or decreases, the function changes. From this fact there arises a question; namely, if the quantity x is increased or decreased, by how much is the function changed, whether it increases or decreases? For the more simple cases, this question is easily answered. If the quantity x is increased by the quantity ω, its square x^2 receives an increase of $2x\omega + \omega^2$. Hence, the increase in x is to the increase of x^2 as ω is to $2x\omega + \omega^2$, that is, as 1 is to $2x + \omega$. In a similar way, we consider the ratio of the increase of x to the increase or decrease that any function of x

receives. Indeed, the investigation of this kind of ratio of increments is not only very important, but it is in fact the foundation of the whole of analysis of the infinite. In order that this may become even clearer, let us take up again the example of the square x^2 with its increment of $2x\omega + \omega^2$, which it receives when x itself is increased by ω. We have seen that the ratio here is $2x + \omega$ to 1. From this it should be perfectly clear that the smaller the increment is taken to be, the closer this ratio comes to the ratio of $2x$ to 1. However, it does not arrive at this ratio before the increment itself, ω, completely vanishes. From this we understand that if the increment of the variable x goes to zero, then the increment of x^2 also vanishes. However, the ratio holds as $2x$ to 1. What we have said here about the square is to be understood of all other functions of x; that is, when their increments vanish as the increment of x vanishes, they have a certain and determinable ratio. In this way, we are led to a definition of *differential calculus*: It is *a method for determining the ratio of the vanishing increments that any functions take on when the variable, of which they are functions, is given a vanishing increment.* It is clearly manifest to those who are not strangers to this subject that the true character of differential calculus is contained in this definition and can be adequately deduced from it.

Therefore, differential calculus is concerned not so much with vanishing increments, which indeed are nothing, but with the ratio and mutual proportion. Since these ratios are expressed as finite quantities, we must think of calculus as being concerned with finite quantities. Although the values seem to be popularly discussed as defined by these vanishing increments, still from a higher point of view, it is always from their ratio that conclusions are deduced. In a similar way, the idea of integral calculus can most conveniently be defined to be *a method for finding those functions from the knowledge of the ratio of their vanishing increments.*

In order that these ratios might be more easily gathered together and represented in calculations, the vanishing increments themselves, although they are really nothing, are still usually represented by certain symbols. Along with these symbols, there is no reason not to give them a certain name. They are called *differentials*, and since they are without quantity, they are also said to be *infinitely small*. Hence, by their nature they are to be so interpreted as absolutely nothing, or they are considered to be equal to nothing. Thus, if the quantity x is given an increment ω, so that it becomes $x + \omega$, its square x^2 becomes $x^2 + 2x\omega + \omega^2$, and it takes the increment $2x\omega + \omega^2$. Hence, the increment of x itself, which is ω, has the ratio to the increment of the square, which is $2x\omega + \omega^2$, as 1 to $2x + \omega$. This ratio reduces to 1 to $2x$, at least when ω vanishes. Let $\omega = 0$, and the ratio of these vanishing increments, which is the main concern of differential calculus, is as 1 to $2x$. On the other hand, this ratio would not be true unless that increment ω vanishes and becomes absolutely equal to zero. Hence, if this nothing that is indicated by ω refers to the increment of the quantity x, since this has the ratio to the increment of the square x^2 as 1 to $2x$, the increment of the square x^2 is equal to $2x\omega$ and for this reason is also equal to zero. Although both of these increments vanish simultaneously, this is no obstacle to their ratios being determined as 1 to $2x$. With respect to this nothing that so far has been represented by the letter ω, in differential calculus we use the symbol dx and call it the differential of x, since it is the increment of the quantity x. When we put dx for ω, the differential of x^2 becomes $2x\,dx$. In a similar way, it is shown that the differential of the cube x^3 will be equal to $3x^2\,dx$. In general, the differential of any quantity x^n will be equal to $nx^{n-1}\,dx$. No matter what other functions of x might be proposed, differential calculus gives rules for finding its differential. Nevertheless, we must constantly keep in mind that since these differentials are absolutely nothing, we can conclude nothing from them except that their mutual ratios reduce to finite quantities. Thus, it is in this way that the principles of differential calculus, which are in agreement with proper reasoning, are established, and all of the objections that are wont to be brought against it crumble spontaneously; but these arguments retain their full rigor if the differentials, that is, the infinitely small, are not completely annihilated.

To many who have discussed the rules of differential calculus, it has seemed that there is a distinction between absolutely nothing and a special order of quantities infinitely small, which do not quite vanish completely but retain a certain quantity that is indeed less than any assignable quantity. Concerning these, it is correctly objected that geometric rigor has been neglected. Because these infinitely small quantities have been neglected, the conclusions that have been drawn are rightly suspected. Although these infinitely small quantities are conceived to be few in number, when even a few, or many, or even an innumerable number of these are neglected, an enormous error may result. There is an attempt wrongfully to refute this objection with examples of this kind, whereby conclusions are drawn from differential calculus in the same way as from elementary geometry. Indeed, if these infinitely small quantities, which are neglected in calculus, are not quite nothing, then necessarily an error must result that will be the greater the more these quantities are heaped up. If it should happen that the error is less, this must be attributed to a fault in the calculation whereby certain errors are compensated by other errors, rather than freeing the calculation from suspicion of error. In order that there be no compensating one error by a new one, let me fix firmly the point I want to make with clear examples. Those quantities that shall be neglected must surely be held to be absolutely nothing. Nor can the infinitely small that is discussed in differential calculus differ in any way from nothing. Even less should this business be ended when the infinitely small is described by some with the example wherein the tiniest mote of dust is compared to a huge mountain or even to the whole terrestrial globe. If someone undertakes to calculate the magnitude of the whole terrestrial globe, it is the custom easily to grant him an error not only of a single grain of dust, but of even many thousands of these. However, geometric rigor shrinks from even so small an error, and this objection would be simply too great were any force granted to it. Then it is difficult to say what possible advantage might be hoped for in distinguishing the infinitely small from absolutely nothing. Perhaps they fear that if they vanish completely, then will be taken away their ratio, to which they feel this whole business leads. It is avowed that it is impossible to conceive how two absolutely nothings can be compared. They think that some magnitude must be left for them that can be compared. They are forced to admit that this magnitude is so small that it is seen as if it were nothing and can be neglected in calculations without error. Neither do they dare to assign any certain and definite magnitude, even though incomprehensibly small. Even if they were assumed to be two or three times smaller, the comparisons are always made in the same way. From this it is clear that this magnitude gives nothing necessary for undertaking a comparison, and so the comparison is not taken away even though that magnitude vanishes completely.

Now, from what has been said above, it is clear that that comparison, which is the concern of differential calculus, would not be valid unless the increments vanish completely. The increment of the quantity x, which we have been symbolizing by ω, has a ratio to the increment of the square x^2, which is $2x\omega + \omega^2$, as 1 to $2x + \omega$. But this always differs from the ratio of 1 to $2x$ unless $\omega = 0$, and if we do require that $\omega = 0$, then we can truly say that this ratio is exactly as 1 to $2x$. In the meantime, it must be understood that the smaller the increment ω becomes, the closer this ratio is approached. It follows that not only is it valid, but quite natural, that these increments be at first considered to be finite and even in drawings, if it is necessary to give illustrations, that they be finitely represented. However, then these increments must be conceived to become continuously smaller, and in this way, their ratio is represented as continuously approaching a certain limit, which is finally attained when the increment becomes absolutely nothing. This limit, which is, as it were, the final ratio of those increments, is the true object of differential calculus. Hence, this ratio must be considered to have laid the very foundation of differential calculus for anyone who has a mind to contemplate these final ratios to which the increments of the variable quantities, as they continuously are more and more diminished, approach and at which they finally arrive.

We find among some ancient authors some trace of these ideas, so that we cannot deny to them at least some conception of the analysis of the infinite. Then gradually this knowledge grew, but it was not all of a sudden that it has arrived at the summit to which it has now come. Even now, there is more that remains obscure than what we see clearly. As differential calculus is extended to all kinds of functions, no matter how they are produced, it is not immediately known what method is to be used to compare the vanishing increments of absolutely all kinds of functions. Gradually this discovery has progressed to more and more complicated functions. For example, for the rational functions, the ultimate ratio that the vanishing increments attain could be assigned long before the time of Newton and Leibniz, so that the differential calculus applied to only these rational functions must be held to have been invented long before that time. However, there is no doubt that Newton must be given credit for that part of differential calculus concerned with irrational functions. This was nicely deduced from his wonderful theorem concerning the general evolution of powers of a binomial. By this outstanding discovery, the limits of differential calculus have been marvelously extended. We are no less indebted to Leibniz insofar as this calculus at that time was viewed as individual tricks, while he put it into the form of a discipline, collected its rules into a system, and gave a crystal-clear explanation. From this there followed great aids in the further development of this calculus, and some of the open questions whose answers were sought were pursued through certain definite principles. Soon, through the studies of both Leibniz and the Bernoullis, the bounds of differential calculus were extended even to transcendental functions, which had in part already been discussed. Then, too, the foundations of integral calculus were firmly established. Those who followed in the elaboration of this field continued to make progress. It was Newton who gave very complete papers in integral calculus, but as to its first discovery, which can hardly be separated from the beginnings of differential calculus, it cannot with absolute certainty be attributed to him. Since the greater part has yet to be developed, it is not possible to say at this time that this calculus has absolutely been discovered. Rather, let us with a grateful mind acknowledge each one according to his efforts toward its completion. This is my judgment as to the attribution of glory for the discovery of this calculus, about which there has been such heated controversy.

No matter what name the mathematicians of different nations are wont to give to this calculus, it all comes to this, that they all agree on this outstanding definition. Whether they call the vanishing increments whose ratios are under consideration by the name differentials or fluxions, these are always understood to be equal to zero, and this must be the true notion of the infinitely small. From this it follows that everything that has been debated about differentials of the second and higher orders, and this has been more out of curiosity then of usefulness, comes back to something very clear, namely, that when everything vanishes together we must consider the mutual ratio rather than the individual quantities. Since the ratio between the vanishing increments of the functions is itself expressed by some function, and if the vanishing increment of this function is compared with others, the result must be considered as the second differential. In this way, we must understand the development of differentials of higher orders, in such a way that they always are seen to be truly finite quantities and that this is the only proper way for them to be represented. At first sight, this description of analysis of the infinite may seem, for the most part, both shallow and extremely sterile, although that obscure notion of the infinitely small hardly offers more. In truth, if the ratios that connect the vanishing increments of any functions are clearly known, then this knowledge very often is of the utmost importance and frequently is so important in extremely arduous investigations that without it almost nothing can be clearly understood. For instance, if the question concerns the motion of a shot fired from a cannon, the air resistance must be known in order to know what the motion will be through a finite distance, as well as both the direction of the path at the beginning and also the velocity, on which the resistance depends. But this changes with time. However, the less distance the shot travels, the less the variation, so that it is possible more easily to come to knowledge of the true relationship. In fact, if we let the distance vanish, since in that case both the difference in direction and change in velocity also are removed, the effect of resistance produced at a single point in time, as well as the change in the path, can be defined exactly. When we know these instantaneous changes or, rather, since these are actually nothing, their mutual relationship, we have gained a great deal. Furthermore, the work of integral calculus is to study changing motion in a finite space. It is my opinion that it is hardly necessary to show further the uses of differential calculus and analysis of the infinite, since it is now sufficiently noted, if even a cursory investigation is made. If we want to study more carefully the motion of either solids or fluids, it cannot be accomplished without analysis of the infinite. Indeed, this science has frequently not been sufficiently cultivated in order that the matter can be accurately explained. Throughout all the branches of mathematics, this higher analysis has penetrated to such an extent that anything that can be explained without its intervention must be esteemed as next to nothing.

I have established in this book the whole of differential calculus, deriving it from true principles and developing it copiously in such a way that nothing pertaining to it that has been discovered so far has been omitted. The work is divided into two parts. In the first part, after laying the foundations of differential calculus, I have presented the method for differentiating every kind of function, for finding not only differentials of the first order, but also those of higher order, and those for functions of a single variable as well as those involving two or more variables. In the second part, I have developed very fully applications of this calculus both in finite analysis and the study of series. In that part, I have also given a very clear explanation of the theorem concerning maxima and minima. As to the application of this calculus to the geometry of plane curves, I have nothing new to offer, and this is all the less to be required, since in other works I have treated this subject so fully. Even with the greatest care, the first principles of differential calculus are hardly sufficiently developed that I should bring them, as it were drawn from geometry, to this science. Here, everything is kept within the bounds of pure analysis, so that in the explanation of the rules of this calculus there is no need for any geometric figures.

Euler

Translator's Introduction

In 1748 Euler published *Introductio in Analysin Infinitorum*, which has been translated as *Introduction to Analysis of the Infinite*, in two books. This can be thought of as Euler's "precalculus." In 1755 he published *Institutiones Calculi Differentialis*. This came in two parts. The first part is the theory of differential calculus, while the second part is concerned with applications of differential calculus. The first part consists of the first nine chapters, with chapters ten through twenty-seven dedicated to the second part. Here, I have translated the first part, that is, the first nine chapters, from Latin into English. The remaining chapters must remain as a future project.

The translation is based on Volume X of the first series of the *Opera Omnia*, edited by Gerhard Kowalewski. I have incorporated in my translation the corrections noted by Kowalewski.

Euler's notation is remarkably modern. However, I have modernized his notation is a few cases. For instance, he rather consistently wrote xx, which I have changed to x^2. For his $\mathrm{l}\, x$, I have written $\ln x$; for $\mathrm{tang}\, x$, $\mathrm{cosc}\, x$, I have written $\tan x$, $\csc x$; and for $\cos x^2$, I have written $\cos^2 x$. I have also modernized his notation for partial derivatives. For his "transcendental quantities depending on a circle," I have substituted "trigonometric quantities."

I would like to thank Kanitra Fletcher, Assistant Editor, and Frank Ganz, TEX Evaluations Manager, at Springer-Verlag New York, Inc., for their generous help. Finally, I would like to thank my wife, Claire, and my children Paul, Drew, and Anne for their patience while I was working on this translation. Special thanks go to my son Jack for his help in the use of the computer.

John D. Blanton

Contents

1

On Finite Differences

1. From what we have said in a previous book[1] about variables and functions, it should be clear enough that as a variable changes, the values of all functions dependent on that variable also change. Thus if a variable quantity x changes by an increment ω, instead of x we write $x + \omega$. Then such functions of x as x^2, x^3, $(a + x) / \left(x^2 + a^2\right)$, take on new values. For instance, x^2 becomes $x^2 + 2x\omega + \omega^2$; x^3 becomes $x^3 + 3x^2\omega + 3x\omega^2 + \omega^3$; $(a + x) / \left(a^2 + x^2\right)$ is transformed into

$$\frac{a + x + \omega}{a^2 + x^2 + 2x\omega + \omega^2}.$$

This kind of change always occurs unless the function has only the appearance of a function of a variable, while in reality it is a constant, for example, x^0. In this case the function remains constant no matter how the value of x changes.

2. Since these things are clear enough, we move now to those results concerning functions upon which rests the whole of analysis of the infinite. Let y be any function of the variable x. Successively we substitute for x the values of an arithmetic progression, that is, x, $x + \omega$, $x + 2\omega$, $x + 3\omega$, $x + 4\omega, \ldots$. We call the value of the function y^{I} when $x + \omega$ is substituted for x; likewise, y^{II} is the value of the function when $x + 2\omega$ is substituted

[1] L. Euler, *Introductio in Analysin Infinitorum.* English translation: *Introduction to Analysis of the Infinite,* Books I, II, Springer-Verlag, New York, 1988.

for x. In a similar way we denote the value of the function by y^{III}, y^{IV}, $y^{\mathrm{V}}, \ldots$, which we obtain when we substitute $x+3\omega$, $x+4\omega$, $x+5\omega, \ldots$. The correspondence between these values is as follows:

$$\begin{array}{ccccccc} x, & x+\omega, & x+2\omega, & x+3\omega, & x+4\omega, & x+5\omega, & \ldots, \\ y, & y^{\mathrm{I}}, & y^{\mathrm{II}}, & y^{\mathrm{III}}, & y^{\mathrm{IV}}, & y^{\mathrm{V}}, & \ldots . \end{array}$$

3. Just as the arithmetic series x, $x+\omega$, $x+2\omega, \ldots$ can be continued to infinity, so the series that depends on the function y: y, y^{I}, $y^{\mathrm{II}}, \ldots$ can be continued to infinity, and its nature will depend on the properties of the function y. Thus if $y=x$ or $y=ax+b$, then the series y, y^{I}, $y^{\mathrm{II}}, \ldots$ is also arithmetic. If $y=a/(bx+c)$, the resulting series will be harmonic. Finally, if $y=a^x$, we will have a geometric series. Furthermore, it is impossible to find any series that does not arise from some such function. We usually call such a function of x, because of the series from which it comes, the *general term* of that series. Since every series formed according to some rule has a general term, so conversely, the series arises from some function of x. This is usually treated at greater length in a discussion of series.

4. Here we will pay special attention to the differences between successive terms of the series y, y^{I}, y^{II}, $y^{\mathrm{III}}, \ldots$. In order that we become familiar with the nature of differentials, we will use the following notation:

$$y^{\mathrm{I}}-y=\Delta y, \quad y^{\mathrm{II}}-y^{\mathrm{I}}=\Delta y^{\mathrm{I}}, \quad y^{\mathrm{III}}-y^{\mathrm{II}}=\Delta y^{\mathrm{II}}, \quad \ldots .$$

We express the increment by Δy, which the function y undergoes when we substitute $x+\omega$ for x, where ω takes any value we wish. In the discussion of series it is usual to take $\omega=1$, but here it is preferable to leave the value general, so that it can be arbitrarily increased or decreased. We usually call this increment Δy of the function y its difference. This is the amount by which the following value y^{I} exceeds the original value y, and we always consider this to be an increment, although frequently it is actually a decrement, since the value may be negative.

5. Since y^{II} is derived from y, if instead of x we write $x+2\omega$, it is clear that we obtain the same result also if we first put $x+\omega$ for x and then again $x+\omega$ for x. It follows that y^{II} is derived from y^{I} if we write $x+\omega$ instead of x. We now see that Δy^{I} is the increment of y^{I} that we obtain when $x+\omega$ is substituted for x. Hence, in like manner, Δy^{I} is called the *difference* of y^{I}. Likewise, Δy^{II} is the *difference* of y^{II}, or its increment, which is obtained by putting $x+\omega$ instead of x. Furthermore, Δy^{III} is the difference, or increment, of y^{III}, and so forth. With this settled, from the series of values of y, namely, y, y^{I}, y^{II}, $y^{\mathrm{III}}, \ldots$, we obtain a series of differences Δy, Δy^{I}, $\Delta y^{\mathrm{II}}, \ldots$, which we find by subtracting each term of the previous series from its successor.

6. Once we have found the series of differences, if we again take the difference of each term and its successor, we obtain a series of differences of differences, which are called *second differences.* We can most conveniently represent these by the following notation:

$$\Delta\Delta y = \Delta y^{\mathrm{I}} - \Delta y,$$
$$\Delta\Delta y^{\mathrm{I}} = \Delta y^{\mathrm{II}} - \Delta y^{\mathrm{I}},$$
$$\Delta\Delta y^{\mathrm{II}} = \Delta y^{\mathrm{III}} - \Delta y^{\mathrm{II}},$$
$$\Delta\Delta y^{\mathrm{III}} = \Delta y^{\mathrm{IV}} - \Delta y^{\mathrm{III}},$$
$$\ldots.$$

We call $\Delta\Delta y$ the second difference of y, $\Delta\Delta y^{\mathrm{I}}$ the second difference of y^{I}, and so forth. In a similar way, from the second differences, if we once more take their differences, we obtain the third differences, which we write as $\Delta^3 y$, $\Delta^3 y^{\mathrm{I}}, \ldots$. Furthermore, we can take the fourth differences $\Delta^4 y$, $\Delta^4 y^{\mathrm{I}}, \ldots$, and even higher, as far as we wish.

7. Let us represent each of these series of differences by the following scheme, in order that we can more easily see their respective relationships:

Arithmetic Progression:

$$x, \quad x+\omega, \quad x+2\omega, \quad x+3\omega, \quad x+4\omega, \quad x+5\omega, \quad \ldots$$

Values of the Function:

$$y, \quad y^{\mathrm{I}}, \quad y^{\mathrm{II}}, \quad y^{\mathrm{III}}, \quad y^{\mathrm{IV}}, \quad y^{\mathrm{V}}, \quad \ldots$$

First Differences:

$$\Delta y, \quad \Delta y^{\mathrm{I}}, \quad \Delta y^{\mathrm{II}}, \quad \Delta y^{\mathrm{III}}, \quad \Delta y^{\mathrm{IV}}, \quad \ldots$$

Second Differences:

$$\Delta\Delta y, \quad \Delta\Delta y^{\mathrm{I}}, \quad \Delta\Delta y^{\mathrm{II}}, \quad \Delta\Delta y^{\mathrm{III}}, \quad \ldots$$

Third Differences:

$$\Delta^3 y, \quad \Delta^3 y^{\mathrm{I}}, \quad \Delta^3 y^{\mathrm{II}}, \quad \ldots$$

Fourth Differences:

$$\Delta^4 y, \quad \Delta^4 y^{\mathrm{I}}, \quad \ldots$$

Fifth Differences:

$$\Delta^5 y, \quad \ldots$$

Each of these series comes from the preceding series by subtracting each term from its successor. Hence, no matter what function of x we substitute for y, it is easy to find each of the series of differences, since the values y^{I}, y^{II}, y^{III},... are easily found from the definition of the function.

8. Let $y = x$, so that $y^{\mathrm{I}} = x^{\mathrm{I}} = x + \omega$, $y^{\mathrm{II}} = x^{\mathrm{II}} = x + 2\omega$, and so forth. When we take the differences, $\Delta x = \omega$, $\Delta x^{\mathrm{I}} = \omega$, $\Delta x^{\mathrm{II}} = \omega, \ldots$, the result is that all of the first differences of x are constant, so that all of the second differences vanish, as do the third differences and all those of higher orders. Since $\Delta x = \omega$, it is convenient to use the notation Δx instead of ω. Since we are assuming that the successive values x, x^{I}, x^{II}, x^{III}, ... form an arithmetic progression, the differences Δx, Δx^{I}, Δx^{II}, ... are constants and mutually equal. It follows that $\Delta\Delta x = 0$, $\Delta^3 x = 0$, $\Delta^4 x = 0$, and so forth.

9. We have assumed that the successive values of x are terms of an arithmetic progression, so that the values of its first differences are constant and its second and succeeding differences vanish. Although the choice is freely ours to make among all possible progressions, still we usually choose the progression to be arithmetic, since it is both the simplest and easiest to understand, and also it has the greatest versatility, in that x can assume absolutely any value. Indeed, if we give ω either negative or positive values in this series, the values of x will always be real numbers. On the other hand, if the series we have chosen is geometric, there is no place for negative values. For this reason the nature of functions y is best determined from the values of x chosen from an arithmetic progression.

10. Just as $\Delta y = y^{\mathrm{I}} - y$, so all the higher differences can also be defined from the terms of the first series: y, y^{I}, y^{II}, y^{III}, ... Since

$$\Delta y^{\mathrm{I}} = y^{\mathrm{II}} - y^{\mathrm{I}},$$

we have

$$\Delta\Delta y = y^{\mathrm{II}} - 2y^{\mathrm{I}} + y$$

and

$$\Delta\Delta y^{\mathrm{I}} = y^{\mathrm{III}} - 2y^{\mathrm{II}} + y^{\mathrm{I}}.$$

Furthermore,

$$\Delta^3 y = \Delta\Delta y^{\mathrm{I}} - \Delta\Delta y = y^{\mathrm{III}} - 3y^{\mathrm{II}} + 3y^{\mathrm{I}} - y;$$

in like manner,

$$\Delta^4 y = y^{\mathrm{IV}} - 4y^{\mathrm{III}} + 6y^{\mathrm{II}} - 4y^{\mathrm{I}} + y$$

and

$$\Delta^5 y = y^{\mathrm{V}} - 5y^{\mathrm{IV}} + 10y^{\mathrm{III}} - 10y^{\mathrm{II}} + 5y^{\mathrm{I}} - y.$$

We observe that the numerical coefficients of these formulas are the same as those of the binomial expansion. Insofar as the first difference is determined by the first two terms of the series y, y^{I}, y^{II}, y^{III}, ..., the second difference is determined by three terms, the third is determined by four terms, and so forth. It follows that when we know the differences of all orders of y, likewise, differences of all orders of y^{I}, y^{II}, ... are defined.

11. It follows that for any function, with any values of x and any differences ω, we can find its first difference as well as its higher differences. Nor is it necessary to compute more terms of the series of the values of y, since we obtain the first difference Δy when for the function y we substitute $x + \omega$ for x and from this value y^{I} we subtract the function y. Likewise the second difference $\Delta\Delta y$ is obtained from the first difference Δy by substituting $x + \omega$ for x to obtain Δy^{I}, and then subtracting Δy from Δy^{I}. In a similar way we get the third difference $\Delta^3 y$ from the second difference $\Delta\Delta y$ by putting $x + \omega$ for x and then subtracting. In the same way we obtain the fourth difference $\Delta^4 y$ and so forth. Provided that we know the first difference of any function, we can find the second, third, and all of the following differences, since the second difference of y is nothing but the first difference of the first difference Δy, and the third difference is nothing but the first difference of the second difference $\Delta\Delta y$, and so forth.

12. If a function y is the sum of two or more functions, as for example $y = p + q + r + \cdots$, then, since $y^{\mathrm{I}} = p^{\mathrm{I}} + q^{\mathrm{I}} + r^{\mathrm{I}} + \cdots$, we have the difference

$$\Delta y = \Delta p + \Delta q + \Delta r + \cdots .$$

Likewise,

$$\Delta\Delta y = \Delta\Delta p + \Delta\Delta q + \Delta\Delta r + \cdots .$$

It follows that if a function is the sum of other functions, then the computation of its differences is just as easy. However, if the function y is the product of two functions p and q, then, since

$$y^{\mathrm{I}} = p^{\mathrm{I}} q^{\mathrm{I}}$$

and

$$p^{\mathrm{I}} = p + \Delta p$$

and

$$q^{\mathrm{I}} = q + \Delta q,$$

we have

$$p^{\mathrm{I}} q^{\mathrm{I}} = pq + p\Delta q + q\Delta p + \Delta p \Delta q,$$

so that

$$\Delta y = p\Delta q + q\Delta p + \Delta p \Delta q.$$

Hence, if p is a constant equal to a, since $\Delta a = 0$ and the function $y = aq$, the first difference Δy equals $a\Delta q$. In a similar way the second difference $\Delta\Delta y$ equals $a\Delta\Delta q$, the third difference $\Delta^3 y$ equals $a\Delta^3 q$, and so forth.

13. Since every polynomial is the sum of several powers of x, we can find all of the differences of polynomials, provided that we know how to find the differences of these powers. For this reason we will investigate the differences of powers of x in the following examples.

Since $x^0 = 1$, we have $\Delta x^0 = 0$, because x^0 does not change when x changes to $x + \omega$.

Also, since as we have seen, $\Delta x = \omega$ and $\Delta\Delta x = 0$, all of the following differences vanish. Since these things are clear, we begin with the second power of x.

Example 1. *Find the differences of all orders of x^2.*

Since here $y = x^2$, we have $y^{\mathrm{I}} = (x + \omega)^2$, so that

$$\Delta y = 2\omega x + \omega^2,$$

and this is the first difference. Now, since ω is a constant, we have $\Delta\Delta y = 2\omega^2$ and $\Delta^3 y = 0$, $\Delta^4 y = 0, \ldots$.

Example 2. *Find the differences of all orders of x^3.*

Let $y = x^3$. Since $y^{\mathrm{I}} = (x + \omega)^3$, we have

$$\Delta y = 3\omega x^2 + 3\omega^2 x + \omega^3,$$

which is the first difference. Then, since $\Delta x^2 = 2\omega x + \omega^2$, we have $\Delta 3\omega x^2 = 6\omega^2 x + 3\omega^3$, $\Delta 3\omega^2 x = 3\omega^3$, and $\Delta\omega^3 = 0$. We put it all together to obtain

$$\Delta\Delta y = 6\omega^2 x + 6\omega^3$$

and

$$\Delta^3 y = 6\omega^3.$$

The differences of higher order vanish.

Example 3. *Find the differences of all orders of* x^4.

Let $y = x^4$. Since $y^{\mathrm{I}} = (x+\omega)^4$, we have

$$\Delta y = 4\omega x^3 + 6\omega^2 x^2 + 4\omega^3 x + \omega^4,$$

which is the first difference. Then, from what we have already found,

$$\Delta 4\omega x^3 = 12\omega^2 x^2 + 12\omega^3 x + 4\omega^4,$$

$$\Delta 6\omega^2 x^2 = 12\omega^3 x + 6\omega^4,$$

$$\Delta 4\omega^3 x = 4\omega^4,$$

$$\Delta \omega^4 = 0.$$

When these are combined, we have the second difference

$$\Delta\Delta y = 12\omega^2 x^2 + 24\omega^3 x + 14\omega^4.$$

Furthermore, since

$$\Delta 12\omega^2 x^2 = 24\omega^3 x + 12\omega^4,$$

$$\Delta 24\omega^3 x = 24\omega^4,$$

$$\Delta 14\omega^4 = 0,$$

we obtain the third difference

$$\Delta^3 y = 24\omega^3 x + 36\omega^4.$$

Finally, we have the fourth difference

$$\Delta^4 y = 24\omega^4,$$

and since this is constant, all differences of higher order vanish.

Example 4. *Find the differences of all orders of* x^n.

Let $y = x^n$. Since $y^{\mathrm{I}} = (x+\omega)^n$, $y^{\mathrm{II}} = (x+2\omega)^n$, $y^{\mathrm{III}} = (x+3\omega)^n, \ldots,$ the expanded powers are as follows:

$$y = x^n,$$

$$y^{\mathrm{I}} = x^n + \frac{n}{1}\omega x^{n-1} + \frac{n(n-1)}{1\cdot 2}\omega^2 x^{n-2} + \frac{n(n-1)(n-2)}{1\cdot 2\cdot 3}\omega^3 x^{n-3} + \cdots,$$

$$y^{\mathrm{II}} = x^n + \frac{n}{1}2\omega x^{n-1} + \frac{n(n-1)}{1\cdot 2}4\omega^2 x^{n-2} + \frac{n(n-1)(n-2)}{1\cdot 2\cdot 3}8\omega^3 x^{n-3} + \cdots,$$

$$y^{\mathrm{III}} = x^n + \frac{n}{1}3\omega x^{n-1} + \frac{n(n-1)}{1\cdot 2}9\omega^2 x^{n-2} + \frac{n(n-1)(n-2)}{1\cdot 2\cdot 3}27\omega^3 x^{n-3} + \cdots,$$

$$y^{\mathrm{IV}} = x^n + \frac{n}{1}4\omega x^{n-1} + \frac{n(n-1)}{1\cdot 2}16\omega^2 x^{n-2} + \frac{n(n-1)(n-2)}{1\cdot 2\cdot 3}64\omega^3 x^{n-3} + \cdots.$$

Then we take the differences to obtain

$$\Delta y = \frac{n}{1}\omega x^{n-1} + \frac{n(n-1)}{1\cdot 2}\omega^2 x^{n-2} + \frac{n(n-1)(n-2)}{1\cdot 2\cdot 3}\omega^3 x^{n-3} + \cdots,$$

$$\Delta y^{\mathrm{I}} = \frac{n}{1}\omega x^{n-1} + \frac{n(n-1)}{1\cdot 2}3\omega^2 x^{n-2} + \frac{n(n-1)(n-2)}{1\cdot 2\cdot 3}7\omega^3 x^{n-3} + \cdots,$$

$$\Delta y^{\mathrm{II}} = \frac{n}{1}\omega x^{n-1} + \frac{n(n-1)}{1\cdot 2}5\omega^2 x^{n-2} + \frac{n(n-1)(n-2)}{1\cdot 2\cdot 3}19\omega^3 x^{n-3} + \cdots,$$

$$\Delta y^{\mathrm{III}} = \frac{n}{1}\omega x^{n-1} + \frac{n(n-1)}{1\cdot 2}7\omega^2 x^{n-2} + \frac{n(n-1)(n-2)}{1\cdot 2\cdot 3}37\omega^3 x^{n-3} + \cdots.$$

Once more we take differences to obtain

$$\Delta\Delta y = n(n-1)\omega^2 x^{n-2} + \frac{n(n-1)(n-2)}{1\cdot 2\cdot 3}6\omega^3 x^{n-3} + \frac{n(n-1)(n-2)(n-3)}{1\cdot 2\cdot 3\cdot 4}14\omega^4 x^{n-4} + \cdots,$$

$$\Delta\Delta y^{\mathrm{I}} = n(n-1)\omega^2 x^{n-2} + \frac{n(n-1)(n-2)}{1\cdot 2\cdot 3}12\omega^3 x^{n-3} + \frac{n(n-1)(n-2)(n-3)}{1\cdot 2\cdot 3\cdot 4}50\omega^4 x^{n-4} + \cdots,$$

$$\Delta\Delta y^{\mathrm{II}} = n(n-1)\omega^2 x^{n-2} + \frac{n(n-1)(n-2)}{1\cdot 2\cdot 3}18\omega^3 x^{n-3} + \frac{n(n-1)(n-2)(n-3)}{1\cdot 2\cdot 3\cdot 4}110\omega^4 x^{n-4} + \cdots.]$$

From these results we use subtraction to derive

$$\Delta^3 y = n(n-1)(n-2)\omega^3 x^{n-3} + \frac{n(n-1)(n-2)(n-3)}{1\cdot 2\cdot 3\cdot 4}36\omega^4 x^{n-4} + \cdots,$$

$$\Delta^3 y^{\mathrm{I}} = n(n-1)(n-2)\omega^3 x^{n-3} + \frac{n(n-1)(n-2)(n-3)}{1\cdot 2\cdot 3\cdot 4}60\omega^4 x^{n-4} + \cdots.$$

Then

$$\Delta^4 y = n(n-1)(n-2)(n-3)\,\omega^4 x^{n-4} + \cdots .$$

14. In order that we may more easily see the law by which these differences of powers of x are formed, let us for the sake of brevity use the following:

$$A = \frac{n}{1},$$
$$B = \frac{n(n-1)}{1 \cdot 2},$$
$$C = \frac{n(n-1)(n-2)}{1 \cdot 2 \cdot 3},$$
$$D = \frac{n(n-1)(n-2)(n-3)}{1 \cdot 2 \cdot 3 \cdot 4},$$
$$E = \frac{n(n-1)(n-2)(n-3)(n-4)}{1 \cdot 2 \cdot 3 \cdot 4 \cdot 5},$$
$$\ldots .$$

We will use the following table for each of the differences:

y	1	0	0	0	0	0	0	0	0	...
Δy	0	1	1	1	1	1	1	1	1	...
$\Delta^2 y$	0	0	2	6	14	30	62	126	254	...
$\Delta^3 y$	0	0	0	6	36	150	540	1,806	5,796	...
$\Delta^4 y$	0	0	0	0	24	240	1,560	8,400	40,824	...
$\Delta^5 y$	0	0	0	0	0	120	1,800	16,800	126,000	...
$\Delta^6 y$	0	0	0	0	0	0	720	15,120	191,520	...
$\Delta^7 y$	0	0	0	0	0	0	0	5,040	141,120	...

Each number in a row of the table is found by taking the sum of the preceding number in that row and the number directly above that preceding number and multiplying that sum by the exponent on Δ. For example, in the row for $\Delta^5 y$ the number 16,800 is found by taking the sum of the preceding 1800 and the 1560 in the preceding row to obtain 3360, which is multiplied by 5.

15. With the aid of this table we can write each of the differences of the powers $y = x^n$ as follows:

$$\Delta y = A\omega x^{n-1} + B\omega^2 x^{n-2} + C\omega^3 x^{n-3} + D\omega^4 x^{n-4} + \cdots,$$

$$\Delta^2 y = 2B\omega^2 x^{n-2} + 6C\omega^3 x^{n-3} + 14D\omega^4 x^{n-4} + \cdots,$$

$$\Delta^3 y = 6C\omega^3 x^{n-3} + 36D\omega^4 x^{n-4} + 150E\omega^5 x^{n-5} + \cdots,$$

$$\Delta^4 y = 24D\omega^4 x^{n-4} + 240E\omega^5 x^{n-5} + 1560F\omega^6 x^{n-6} + \cdots.$$

In general, the difference of order m of the power x^n, that is $\Delta^m y$, is expressed in the following way.

Let

$$I = \frac{n(n-1)(n-2)\cdots(n-m+1)}{1\cdot 2\cdot 3\cdots m},$$

$$K = \frac{n-m}{m+1} I,$$

$$L = \frac{n-m-1}{m+2} K,$$

$$M = \frac{n-m-2}{m+3} L,$$

$$\cdots.$$

Then we let

$$\alpha = (m+1)^m - \frac{m}{1} m^m + \frac{m(m-1)}{1\cdot 2}(m-1)^m - \frac{m(m-1)(m-2)}{1\cdot 2\cdot 3}(m-2)^m + \cdots,$$

$$\beta = (m+1)^{m+1} - \frac{m}{1} m^{m+1} + \frac{m(m-2)}{1\cdot 2}(m-1)^{m+1} - \frac{m(m-1)(m-2)}{1\cdot 2\cdot 3}(m-2)^{m+1} + \cdots,$$

$$\gamma = (m+1)^{m+2} - \frac{m}{1} m^{m+2} + \frac{m(m-1)}{1\cdot 2}(m-1)^{m+2} - \frac{m(m-1)(m-2)}{1\cdot 2\cdot 3}(m-2)^{m+2} + \cdots.$$

With these definitions we can write

$$\Delta^m y = \alpha I\omega^m x^{n-m} + \beta K\omega^{m+1} x^{n-m-1} + \gamma L\omega^{m+2} x^{n-m-2} + \cdots.$$

This result follows immediately from all of the differences of y, y^{I}, y^{II}, $y^{\mathrm{III}}, \ldots$.

16. From what we have seen it is clear that if the exponent n is a positive integer, sooner or later we obtain a constant difference, and thereafter all differences vanish. Thus we have

$$\Delta.x = \omega,$$

$$\Delta^2.x^2 = 2\omega^2,$$

$$\Delta^3.x^3 = 6\omega^3,$$

$$\Delta^4.x^4 = 24\omega^4,$$

and finally,

$$\Delta^n.x^n = 1 \cdot 2 \cdot 3 \cdots n\omega^n$$

(see paragraph 146 for an explanation of this notation). It follows that every polynomial finally arrives at a constant difference. For instance, the linear function of x, $ax+b$, has for a first difference the constant $a\omega$. The quadratic function $ax^2 + bx + c$ has for second difference the constant $2a\omega^2$. A third-degree polynomial has its third difference constant; the fourth degree has its fourth difference constant, and so forth.

17. The method whereby we find the differences of powers x^n can be further extended to exponents that are negative, a fraction, or even an irrational number. For the sake of clarity we will discuss only the first differences of powers with these kinds of exponents, since the law for second and higher differences is not so easily seen. Let

$$\Delta.x = \omega,$$

$$\Delta.x^2 = 2\omega x + \omega^2,$$

$$\Delta.x^3 = 3\omega x^2 + 3\omega^2 x + \omega^3,$$

$$\Delta.x^4 = 4\omega x^3 + 6\omega^2 x^2 + 4\omega^3 x + \omega^4,$$

$$\ldots$$

In a similar way we let

$$\Delta.x^{-1} = -\frac{\omega}{x^2} + \frac{\omega^2}{x^3} - \frac{\omega^3}{x^4} + \cdots,$$

$$\Delta.x^{-2} = -\frac{2\omega}{x^3} + \frac{3\omega^2}{x^4} - \frac{4\omega^3}{x^5} + \cdots,$$

$$\Delta.x^{-3} = -\frac{3\omega}{x^4} + \frac{6\omega^2}{x^5} - \frac{10\omega^3}{x^6} + \cdots,$$

$$\Delta.x^{-4} = -\frac{4\omega}{x^5} + \frac{10\omega^2}{x^6} - \frac{20\omega^3}{x^7} + \cdots.$$

We continue in the same way for the rest. For fractions we have

$$\Delta.x^{1/2} = \frac{\omega}{2x^{1/2}} - \frac{\omega^2}{8x^{3/2}} + \frac{\omega^3}{16x^{5/2}} - \cdots,$$

$$\Delta.x^{1/3} = \frac{\omega}{3x^{2/3}} - \frac{\omega^2}{9x^{5/9}} + \frac{5\omega^3}{81x^{8/3}} - \cdots,$$

$$\Delta.x^{-1/2} = -\frac{\omega}{2x^{3/2}} + \frac{3\omega^2}{8x^{5/2}} - \frac{5\omega^3}{16x^{7/2}} + \cdots,$$

$$\Delta.x^{-1/3} = -\frac{\omega}{3x^{4/3}} + \frac{2\omega^2}{9x^{7/3}} - \frac{14\omega^3}{81x^{10/3}} + \cdots.$$

18. It should be clear that if the exponent is not a positive integer, then these differences will progress without limit, that is, there will be an infinite number of terms. Nevertheless, these same differences can be expressed by a finite expression. If we let $y = x^{-1} = 1/x$, then $y^{\mathrm{I}} = 1/(x+\omega)$, so that

$$\Delta.x^{-1} = \Delta.\frac{1}{x} = \frac{1}{x+\omega} - \frac{1}{x}.$$

Hence, if the fraction $1/(x+\omega)$ is expressed as a series, then we obtain the infinite expression we saw before. In a similar way we have

$$\Delta.x^{-2} = \Delta.\frac{1}{x^2} = \frac{1}{(x+\omega)^2} - \frac{1}{x^2}.$$

Furthermore, for irrational expressions we have

$$\Delta.\sqrt{x} = \sqrt{x+\omega} - \sqrt{x}$$

and

$$\Delta.\frac{1}{\sqrt{x}} = \frac{1}{\sqrt{x+\omega}} - \frac{1}{\sqrt{x}}.$$

If these formulas are expressed as series in the usual way, we will obtain the expressions found above.

19. In this same way, differences of functions, either rational or irrational, can be found. If, for example, we wish to find the first difference of the fraction $1/\left(a^2+x^2\right)$, then we let $y = 1/\left(a^2+x^2\right)$, and since

$$y^{\mathrm{I}} = \frac{1}{a^2+x^2+2\omega x+\omega^2},$$

we have

$$\Delta y = \Delta \frac{1}{a^2+x^2} = \frac{1}{a^2+x^2+2\omega x+\omega^2} - \frac{1}{a^2+x^2},$$

and this expression can be converted into an infinite series.

We let $a^2+x^2 = P$ and $2\omega x+\omega^2 = Q$. Then

$$\frac{1}{P+Q} = \frac{1}{P} - \frac{Q}{P^2} + \frac{Q^2}{P^3} - \frac{Q^3}{P^4} + \cdots$$

and

$$\Delta y = -\frac{Q}{P^2} + \frac{Q^2}{P^3} - \frac{Q^3}{P^4} + \cdots .$$

When we substitute the values of P and Q we obtain

$$\begin{aligned} \Delta y &= \Delta \frac{1}{a^2+x^2} \\ &= -\frac{2\omega x+\omega^2}{\left(a^2+x^2\right)^2} + \frac{4\omega^2x^2+4\omega^3x+\omega^4}{\left(a^2+x^2\right)^3} \\ &\quad - \frac{8\omega^3x^3+12\omega^4x^2+6\omega^5x+\omega^6}{\left(a^2+x^2\right)^4} + \cdots . \end{aligned}$$

If these terms are ordered by the powers of ω, we obtain

$$\Delta . \frac{1}{x^2+a^2} = -\frac{2\omega x}{\left(a^2+x^2\right)^2} + \frac{\omega^2\left(3x^2-a^2\right)}{\left(a^2+x^2\right)^3} - \frac{4\omega^3\left(x^3-a^2x\right)}{\left(a^2+x^2\right)^4} + \cdots .$$

20. Differences of irrational functions can be expressed by similar series.

If we let $y = \sqrt{a^2+x^2}$, and since

$$y^{\mathrm{I}} = \sqrt{a^2+x^2+2\omega x+\omega^2},$$

we let $a^2+x^2 = P$ and $2\omega x+\omega^2 = Q$, then

$$\Delta y = \sqrt{P+Q} - \sqrt{P} = \frac{Q}{2\sqrt{P}} - \frac{Q^2}{8P\sqrt{P}} + \frac{Q^3}{16P^2\sqrt{P}} - \cdots,$$

so that

$$\Delta y = \Delta.\sqrt{a^2+x^2} = \frac{2\omega x+\omega^2}{2\sqrt{a^2+x^2}} - \frac{4\omega^2x^2+4\omega^3x+\omega^4}{8\left(a^2+x^2\right)\sqrt{a^2+x^2}} + \cdots,$$

or

$$\Delta y = \frac{\omega x}{\sqrt{a^2+x^2}} + \frac{a^2\omega^2}{2\left(a^2+x^2\right)\sqrt{a^2+x^2}} - \frac{a^2\omega^3 x}{2\left(a^2+x^2\right)^2\sqrt{a^2+x^2}} + \cdots.$$

From this we gather the fact that the difference of any function of x, which we call y, can be put into this form, so that

$$\Delta y = P\omega + Q\omega^2 + R\omega^3 + S\omega^4 + \cdots,$$

where P, Q, R, $S, \ldots$ are certain functions of x that in any case can be defined in terms of the function y.

21. We do not exclude from this form of expression even the differences of transcendental functions, as will clearly appear from the following examples.

Example 1. *Find the first difference of the natural logarithm of* x.

Let $y = \ln x$. Since $y^{\mathrm{I}} = \ln(x+\omega)$, we have

$$\Delta y = y^{\mathrm{I}} - y = \ln(x+\omega) - \ln x = \ln\left(1+\frac{\omega}{x}\right).$$

Elsewhere[2] we have shown how this kind of logarithm can be expressed in an infinite series. We use this to obtain

$$\Delta y = \Delta \ln x = \frac{\omega}{x} - \frac{\omega^2}{2x^2} + \frac{\omega^3}{3x^3} - \frac{\omega^4}{4x^4} + \cdots.$$

Example 2. *Find the first difference of exponential functions* a^x.

Let $y = a^x$, so that $y^{\mathrm{I}} = a^x a^\omega$. We have also shown[3] that

$$a^\omega = 1 + \frac{\omega \ln a}{1} + \frac{\omega^2(\ln a)^2}{1\cdot 2} + \frac{\omega^3(\ln a)^3}{1\cdot 2\cdot 3} + \cdots.$$

From this we have

$$\Delta.a^x = y^{\mathrm{I}} - y = \Delta y = \frac{a^x\omega\ln a}{1} + \frac{a^x\omega^2(\ln a)^2}{1\cdot 2} + \frac{a^x\omega^3(\ln a)^3}{1\cdot 2\cdot 3} + \cdots.$$

Example 3. *In a unit circle, to find the difference of the sine of the arc* x.

---

[2] *Introduction to Analysis of the Infinite,* Book I, Chapter VII; see also note on page 1.
[3] *Introduction,* Book I, Chapter VIII; see also note on page 1.

Let $\sin x = y$. Then $y^{\mathrm{I}} = \sin(x+\omega)$, so that

$$\Delta y = y^{\mathrm{I}} - y = \sin(x+\omega) - \sin x.$$

Now,

$$\sin(x+\omega) = \cos\omega \cdot \sin x + \sin\omega \cdot \cos x,$$

and we have shown[4] that

$$\cos\omega = 1 - \frac{\omega^2}{1\cdot 2} + \frac{\omega^4}{1\cdot 2\cdot 3\cdot 4} - \frac{\omega^6}{1\cdot 2\cdot 3\cdot 4\cdot 5\cdot 6} + \cdots$$

and

$$\sin\omega = \omega - \frac{\omega^3}{1\cdot 2\cdot 3} + \frac{\omega^5}{1\cdot 2\cdot 3\cdot 4\cdot 5} - \frac{\omega^7}{1\cdot 2\cdot 3\cdot 4\cdot 5\cdot 6\cdot 7} + \cdots.$$

When we substitute these series we obtain

$$\Delta.\sin x = \omega\cos x - \frac{\omega^2}{2}\sin x - \frac{\omega^3}{6}\cos x + \frac{\omega^4}{24}\sin x + \frac{\omega^5}{120}\cos x - \cdots.$$

Example 4. *In a unit circle, to find the difference of the cosine of the arc* x.

Let $y = \cos x$. Then since $y^{\mathrm{I}} = \cos(x+\omega)$, we have

$$y^{\mathrm{I}} = \cos\omega\cos x - \sin\omega\sin x$$

and

$$\Delta y = \cos\omega\cos x - \sin\omega\sin x - \cos x.$$

From the series referenced above we obtain

$$\Delta.\cos x = -\omega\sin x - \frac{\omega^2}{2}\cos x - \frac{\omega^3}{6}\sin x + \frac{\omega^4}{24}\cos x - \frac{\omega^5}{120}\sin x - \cdots.$$

22. Since any function of x, which we call y, whether it is algebraic or transcendental, has a difference of the form

$$\Delta y = P\omega + Q\omega^2 + R\omega^3 + S\omega^4 + \cdots,$$

if we take the difference again, it is clear that the second difference of y has the form

$$\Delta^2 y = P\omega^2 + Q\omega^3 + R\omega^4 + \cdots.$$

---

[4] *Introduction*, Book I, Chapter VIII; see also note on page 1.

In a similar way the third difference will be

$$\Delta^3 y = P\omega^3 + Q\omega^4 + R\omega^5 + \cdots,$$

and so forth.

We should note that these letters P, Q, $R, \ldots$ do not stand for determined values, nor does the same letter in different differences denote the same function of x. Indeed, we use the same letters lest we run out of symbols.

Furthermore, these forms of differences should be carefully noted, since they are very useful in the analysis of the infinite.

23. According to the method we are using, the first difference of any function is found, and from it we find the differences of the successive orders. Indeed, from the values of successive functions of y, namely, y^{I}, y^{II}, y^{III}, $y^{\mathrm{IV}}, \ldots$, we find in turn differences of y of any order. We recall that

$$\begin{aligned} y^{\mathrm{I}} &= y + \Delta y, \\ y^{\mathrm{II}} &= y + 2\Delta y + \Delta^2 y, \\ y^{\mathrm{III}} &= y + 3\Delta y + 3\Delta^2 y + \Delta^3 y, \\ y^{\mathrm{IV}} &= y + 4\Delta y + 6\Delta^2 y + 4\Delta^3 y + \Delta^4 y, \end{aligned}$$

and so forth, where the coefficients arise from the binomial expansion. Since y^{I}, y^{II}, $y^{\mathrm{III}}, \ldots$ are values of y that arise when we substitute for x the successive values $x + \omega$, $x + 2\omega$, $x + 3\omega, \ldots$, we can immediately assign the value of $y^{(n)}$, which is produced if in place of x we write $x + n\omega$. The value obtained is

$$y + \frac{n}{1}\Delta y + \frac{n(n-1)}{1 \cdot 2}\Delta^2 y + \frac{n(n-1)(n-2)}{1 \cdot 2 \cdot 3}\Delta^3 y + \cdots.$$

Furthermore, values of y can be obtained even if n is a negative integer. Thus, if instead of x we put $x - \omega$, the function y is in the form

$$y - \Delta y + \Delta^2 y - \Delta^3 y + \Delta^4 y - \cdots.$$

If instead of x we put $x - 2\omega$, the function y becomes

$$y - 2\Delta y + 3\Delta^2 y - 4\Delta^3 y + 5\Delta^4 y - \cdots.$$

24. We will add a few things about the inverse problem. That is, if we are given the difference of some function, we would like to investigate the function itself. Since this is generally very difficult and frequently requires analysis of the infinite, we will discuss only some of the easier cases. First of all, proceeding backwards, if we have found the difference for some function and that difference is now given, we can, in turn, exhibit that function from which the difference came. Thus, since the difference of the function $ax+b$ is $a\omega$, if we are asked for the function whose difference is $a\omega$, we can immediately reply that the function is $ax+b$, since the constant quantity b does not appear in the difference, so we are free to choose any value for b. It is always the case that if the difference of a function P is Q, then the function $P+A$, where A is any constant, also has Q as its difference. It follows that if this difference Q is given, a function from which this came is $P+A$. Since A is arbitrary, the function does not have a determined value.

25. We call that desired function, whose difference is given, the *sum*. This name is appropriate, since a sum is the operation inverse to difference, but also since the desired function really is the sum of all of the antecedent values of the difference. Just as

$$y^{\mathrm{I}} = y + \Delta y$$

and

$$y^{\mathrm{II}} = y + \Delta y + \Delta y^{\mathrm{I}},$$

if the values of y are continued backwards in such a way that what $x-\omega$ corresponds to is written as y_{I}, and y_{II} preceding this, and also y_{III}, y_{IV}, $y_{\mathrm{V}}, \ldots$, and if we form the retrograde series with their differences

$$y_{\mathrm{V}}, \qquad y_{\mathrm{IV}}, \qquad y_{\mathrm{III}}, \qquad y_{\mathrm{II}}, \qquad y_{\mathrm{I}}, \qquad y$$

and

$$\Delta y_{\mathrm{V}}, \qquad \Delta y_{\mathrm{IV}}, \qquad \Delta y_{\mathrm{III}}, \qquad \Delta y_{\mathrm{II}}, \qquad \Delta y_{\mathrm{I}},$$

then $y = \Delta y_{\mathrm{I}} + y_{\mathrm{I}}$. Since $y_{\mathrm{I}} = \Delta y_{\mathrm{II}} + y_{\mathrm{II}}$ and $y_{\mathrm{II}} = \Delta y_{\mathrm{III}} + y_{\mathrm{III}}$, we have

$$y = \Delta y_{\mathrm{I}} + \Delta y_{\mathrm{II}} + \Delta y_{\mathrm{III}} + \Delta y_{\mathrm{IV}} + \Delta y_{\mathrm{V}} + \cdots.$$

Thus the function y, whose difference is Δy, is the sum of the values of the antecedent differences, which we obtain when instead of x we write the antecedent values $x-\omega$, $x-2\omega$, $x-3\omega$, ...

26. Just as we used the symbol Δ to signify a difference, so we use the symbol Σ to indicate a sum. For example, if z is the difference of the function y, then $\Delta y = z$. We have previously discussed how to find the difference z if y is given. However, if z is given and we want to find its sum y, we let $y = \Sigma z$, and from the equation $z = \Delta y$, working backwards, we obtain the equation $y = \Sigma z$, where an arbitrary constant can be added for the reason already discussed. From the equation $z = \Delta y$, if we invert, we also obtain $y = \Sigma z + C$. Now, since the difference of ay is $a\Delta y = az$, we have $\Sigma az = ay$, provided that a is a constant. Since $\Delta x = \omega$, we have $\Sigma\omega = x + C$ and $\Sigma a\omega = ax + C$; since ω is a constant, we have $\Sigma\omega^2 = \omega x + C$, $\Sigma\omega^3 = \omega^2 + C$, and so forth.

27. If we invert the differences of powers of x which we previously found, we have $\Sigma\omega = x$ and from this $\Sigma 1 = x/\omega$. Then we have

$$\Sigma\left(2\omega x + \omega^2\right) = x^2,$$

so that

$$\Sigma x = \frac{x^2}{2\omega} - \Sigma\frac{\omega}{2} = \frac{x^2}{2\omega} - \frac{x}{2}.$$

Furthermore,

$$\Sigma\left(3\omega x^2 + 3\omega^2 x + \omega^3\right) = x^3,$$

or

$$3\omega\Sigma x^2 + 3\omega^2\Sigma x + \omega^3\Sigma 1 = x^3,$$

so that

$$\Sigma x^2 = \frac{x^3}{3\omega} - \omega\Sigma x - \frac{\omega^2}{3}\Sigma 1,$$

and so

$$\Sigma x^2 = \frac{x^3}{3\omega} - \frac{x^2}{2} + \frac{\omega x}{6}.$$

In a similar way we have

$$\Sigma x^3 = \frac{x^4}{4\omega} - \frac{3\omega}{2}\Sigma x^2 - \omega^2\Sigma x - \frac{\omega^3}{4}\Sigma 1.$$

If for Σx^2, Σx, and $\Sigma 1$ we substitute the previously found values, we obtain

$$\Sigma x^3 = \frac{x^4}{4\omega} - \frac{x^3}{2} + \frac{\omega x^2}{4}.$$

Then, since

$$\Sigma x^4 = \frac{x^5}{5\omega} - 2\omega\Sigma x^3 - 2\omega\Sigma x^2 - \omega^3\Sigma x - \frac{\omega^4}{5}\Sigma 1,$$

when we make the appropriate substitutions we have

$$\Sigma x^4 = \frac{x^5}{5\omega} - \frac{1}{2}x^4 + \frac{1}{3}\omega x^3 - \frac{1}{30}\omega^3 x.$$

In a similar way we obtain

$$\Sigma x^5 = \frac{x^6}{6\omega} - \frac{1}{2}x^5 + \frac{5}{12}\omega x^4 - \frac{1}{12}\omega^3 x^4$$

and

$$\Sigma x^6 = \frac{x^7}{7\omega} - \frac{1}{2}x^6 + \frac{1}{2}\omega x^5 - \frac{1}{6}\omega^3 x^3 - \frac{1}{42}\omega^5 x.$$

Later we will show an easier method to obtain these expressions.

28. If the given difference is for a polynomial function of x, then its sum (or the function of which it is the difference) can easily be found with these formulas. Since the difference is made up of different powers of x, we find the sum of each term and then collect all of these terms.

Example 1. *Find the function whose difference is* $ax^2 + bx + c$.

We find the sum of each term by means of the formulas found above:

$$\Sigma ax^2 = \frac{ax^3}{3\omega} - \frac{ax^2}{2} + \frac{a\omega x}{6},$$

$$\Sigma bx = \frac{bx^2}{2\omega} - \frac{bx}{2},$$

$$\Sigma c = \frac{cx}{\omega}.$$

When we collect these sums we obtain

$$\Sigma\left(ax^2 + bx + c\right) = \frac{ax^3}{3\omega} - \frac{a\omega - b}{2\omega}x^2 + \frac{a\omega^2 - 3b\omega + 6c}{6\omega}x + C,$$

which is the desired function, whose difference is $ax^2 + bx + c$.

Example 2. *Find the function whose difference is* $x^4 - 2\omega^2 x^2 + \omega^4$.

Following the same method we obtain

$$\Sigma x^4 = \frac{1}{5\omega}x^5 - \frac{1}{2}x^4 + \frac{\omega x^3}{3} - \frac{\omega^3}{30}x,$$

$$-\Sigma 2\omega^2 x^2 = -\frac{2\omega}{3}x^3 + \omega^2 x^2 - \frac{\omega^3}{3}x,$$

$$+\Sigma\omega^4 = \omega^3 x,$$

so that the desired function is

$$\frac{1}{5\omega}x^5 - \frac{1}{2}x^4 - \frac{1}{3}\omega x^3 + \omega^2 x^2 + \frac{19}{30}\omega^3 x + C.$$

As a check, if instead of x we put $x+\omega$ and from this expression we subtract the one we have found, the given difference $x^4 - 2\omega^2 x^2 + \omega^4$ is what remains.

29. If we carefully observe the sums of the powers of x that we have found, the first, second, and third terms, we quickly discover the laws of formation that they follow. The law for the following terms is not so obvious that we can state in general the sum for the power x^n. Later (in paragraph 132 of the second part) we will show that

$$\begin{aligned}
\Sigma x^n = {} & \frac{x^{n+1}}{(n+1)\,\omega} - \frac{1}{2}x^n + \frac{1}{2}\cdot\frac{n\omega}{2\cdot 3}x^{n-1} - \frac{1}{6}\cdot\frac{n\,(n-1)\,(n-2)\,\omega^3}{2\cdot 3\cdot 4\cdot 5}x^{n-3} \\
& + \frac{1}{6}\cdot\frac{n\,(n-1)\,(n-2)\,(n-3)\,(n-4)\,\omega^5}{2\cdot 3\cdot 4\cdot 5\cdot 6\cdot 7}x^{n-5} \\
& - \frac{3}{10}\cdot\frac{n\,(n-1)\cdots(n-6)\,\omega^7}{2\cdot 3\cdots 8\cdot 9}x^{n-7} \\
& + \frac{5}{6}\cdot\frac{n\,(n-1)\cdots(n-8)\,\omega^9}{2\cdot 3\cdots 10\cdot 11}x^{n-9} \\
& - \frac{691}{210}\cdot\frac{n\,(n-1)\cdots(n-10)\,\omega^{11}}{2\cdot 3\cdots 12\cdot 13}x^{n-11} \\
& + \frac{35}{2}\cdot\frac{n\,(n-1)\cdots(n-12)\,\omega^{13}}{2\cdot 3\cdots 14\cdot 15}x^{n-13} \\
& - \frac{3617}{30}\cdot\frac{n\,(n-1)\cdots(n-14)\,\omega^{15}}{2\cdot 3\cdots 16\cdot 17}x^{n-15} \\
& + \frac{43867}{42}\cdot\frac{n\,(n-1)\cdots(n-16)\,\omega^{17}}{2\cdot 3\cdots 18\cdot 19}x^{n-17} \\
& - \frac{1222277}{110}\cdot\frac{n\,(n-1)\cdots(n-18)\,\omega^{19}}{2\cdot 3\cdots 20\cdot 21}x^{n-19} \\
& + \frac{854513}{6}\cdot\frac{n\,(n-1)\cdots(n-20)\,\omega^{21}}{2\cdot 3\cdots 22\cdot 23}x^{n-21} \\
& - \frac{1181820455}{546}\cdot\frac{n\,(n-1)\cdots(n-22)\,\omega^{23}}{2\cdot 3\cdots 24\cdot 25}x^{n-23} \\
& + \frac{76977927}{2}\cdot\frac{n\,(n-1)\cdots(n-24)\,\omega^{25}}{2\cdot 3\cdots 26\cdot 27}x^{n-25} \\
& - \frac{23749461029}{30}\cdot\frac{n\,(n-1)\cdots(n-26)\,\omega^{27}}{2\cdot 3\cdots 28\cdot 29}x^{n-27} \\
& + \frac{8615841276005}{462}\cdot\frac{n\,(n-1)\cdots(n-28)\,\omega^{29}}{2\cdot 3\cdots 30\cdot 31}x^{n-29} \\
& + \cdots + C.
\end{aligned}$$

The main interest here is the sequence of purely numerical coefficients. It is not yet time to explain how these are formed.

30. It is clear that if n is not a positive integer, then the expression for the sum is going to be an infinite series, nor can it be expressed in finite form. Furthermore, here we should note that not all powers of x with exponents less than n occur. All of the terms x^{n-2}, x^{n-4}, $x^{n-6}, \ldots$ are lacking, that is, they have coefficients equal to zero, although the second term, x^n, does not follow this law, since it has coefficient $-\frac{1}{2}$. If n is negative or a fraction, then this sum can be expressed as an infinite series with the sole exception that n cannot be -1, since in that case the term

$$\frac{x^{n+1}}{(n+1)\,\omega}$$

would be infinite, since $n+1=0$. Hence, if $n=-2$, then

$$\Sigma\frac{1}{x^2} = C - \frac{1}{\omega x} - \frac{1}{2x^2} - \frac{1}{2}\cdot\frac{\omega}{3x^3} + \frac{1}{6}\cdot\frac{\omega^3}{5x^5} - \frac{1}{6}\cdot\frac{\omega^5}{7x^7} + \frac{3}{10}\cdot\frac{\omega^7}{9x^9}$$
$$- \frac{5}{6}\cdot\frac{\omega^9}{11x^{11}} + \frac{691}{210}\cdot\frac{\omega^{11}}{13x^{13}} - \frac{35}{2}\cdot\frac{\omega^{13}}{15x^{15}} + \frac{3617}{30}\cdot\frac{\omega^{15}}{17x^{17}} - \cdots .$$

31. If a given difference is any power of x, then its sum, or the function from which it came, can be given. However, if the given difference is of some other form, so that it cannot be expressed in parts that are powers of x, then the sum may be very difficult, and frequently impossible, to find, unless by chance it is clear that it came from some function. For this reason it is useful to investigate the difference of many functions and carefully to note them, so that when this difference is given, its sum or the function from which it came can be immediately given. In the meantime, the method of infinite series will supply many rules whose use will marvelously aid in finding sums.

32. Frequently, it is easier to find the sum if the given difference can be expressed as a product of linear factors that form an arithmetic progression whose difference is ω. Suppose the given function is $(x+\omega)(x+2\omega)$. Since when we substitute $x+\omega$ for x we obtain $(x+2\omega)(x+3\omega)$, then the difference will be $2\omega(x+2\omega)$. Hence, going backwards, if the given difference is $2\omega(x+2\omega)$, then its sum is $(x+\omega)(x+2\omega)$. From this it follows that

$$\Sigma(x+2\omega) = \frac{1}{2\omega}(x+\omega)(x+2\omega).$$

Similarly, if the given function is $(x+n\omega)(x+(n+1)\omega)$, since its difference is $2\omega(x+(n+1)\omega)$, we have

$$\Sigma(x+(n+1)\omega) = \frac{1}{2\omega}(x+n\omega)(x+(n+1)\omega),$$

and

$$\Sigma\,(x+n\omega)=\frac{1}{2\omega}\,(x+(n-1)\,\omega)\,(x+n\omega)\,.$$

33. If the function is the product of several factors, such as

$$y=(x+(n-1)\,\omega)\,(x+n\omega)\,(x+(n+1)\,\omega)\,,$$

then since

$$y^{\mathrm{I}}=(x+n\omega)\,(x+(n+1)\,\omega)\,(x+(n+2)\,\omega)\,,$$

we have

$$\Delta y=3\omega\,(x+n\omega)\,(x+(n+1)\,\omega)\,.$$

It follows that

$$\Sigma\,(x+n\omega)\,(x+(n+1)\,\omega)=\frac{1}{3\omega}\,(x+(n-1)\,\omega)\,(x+n\omega)\,(x+(n+1)\,\omega)\,.$$

In the same way we find that

$$\Sigma\,(x+n\omega)\,(x+(n+1)\,\omega)\,(x+(n+2)\,\omega)$$
$$=\frac{1}{4\omega}\,(x+(n-1)\,\omega)\,(x+n\omega)\,(x+(n+1)\,\omega)\,(x+(n+2)\,\omega)\,.$$

Hence the law for finding sums is quite clear if the difference is the product of several factors of this kind. Although these differences are polynomials, still this method of finding their sums seems to be easier than the previous method.

34. From this method the way is now clear to finding the sums of fractions. Let the given fraction be

$$y=\frac{1}{x+n\omega}.$$

Since

$$y^{\mathrm{I}}=\frac{1}{x+(n+1)\,\omega},$$

we have

$$\Delta y=\frac{1}{x+(n+1)\,\omega}-\frac{1}{x+n\omega}=\frac{-\omega}{(x+n\omega)\,(x+(n+1)\,\omega)},$$

and it follows that

$$\Sigma\frac{1}{(x+n\omega)\,(x+(n+1)\,\omega)}=-\frac{1}{\omega}\cdot\frac{1}{x+n\omega}.$$

Furthermore, let

$$y = \frac{1}{(x+n\omega)(x+(n+1)\omega)}.$$

Since

$$y^{\mathrm{I}} = \frac{1}{(x+(n+1)\omega)(x+(n+2)\omega)},$$

we have

$$\Delta y = \frac{-2\omega}{(x+n\omega)(x+(n+1)\omega)(x+(n+2)\omega)},$$

and it follows that

$$\Sigma\frac{1}{(x+n\omega)(x+(n+1)\omega)(x+(n+2)\omega)}$$
$$= -\frac{1}{2\omega}\cdot\frac{1}{(x+n\omega)(x+(n+1)\omega)}.$$

In a similar way we have

$$\Sigma\frac{1}{(x+n\omega)(x+(n+1)\omega)(x+(n+2)\omega)(x+(n+3)\omega)}$$
$$= -\frac{1}{3\omega}\cdot\frac{1}{(x+n\omega)(x+(n+1)\omega)(x+(n+2)\omega)}.$$

35. We should observe this method carefully, since sums of differences of this kind cannot be found by the previous method. If the difference has a numerator or the denominator has factors that do not form an arithmetic progression, then the safest method for finding sums is to express the fraction as the sum of partial fractions. Although we may not be able to find the sum of an individual fraction, it may be possible to consider them in pairs. We have only to see whether it may be possible to use the formula

$$\Sigma\frac{1}{x+(n+1)\omega} - \Sigma\frac{1}{x+n\omega} = \frac{1}{x+n\omega}.$$

Although neither of these sums is known, still their difference is known.

36. In these cases the problem is reduced to finding the partial fractions, and this is treated at length in a previous book.[5] In order that we may see its usefulness for finding sums, we will consider some examples.

[5] *Introduction*, Book I, Chapter II.

Example 1. *Find the function whose difference is*

$$\frac{3x+2\omega}{x\,(x+\omega)\,(x+2\omega)}.$$

The given difference is expressed as partial fractions:

$$\frac{1}{\omega}\cdot\frac{1}{x}+\frac{1}{\omega}\cdot\frac{1}{x+\omega}-\frac{2}{\omega}\cdot\frac{1}{x+2\omega}.$$

From the previous formula

$$\Sigma\frac{1}{x+n\omega}=\Sigma\frac{1}{x+(n+1)\,\omega}-\frac{1}{x+n\omega},$$

we have

$$\Sigma\frac{1}{x}=\Sigma\frac{1}{x+\omega}-\frac{1}{x}.$$

It follows that the desired sum is

$$\frac{1}{\omega}\Sigma\frac{1}{x}+\frac{1}{\omega}\Sigma\frac{1}{x+\omega}-\frac{2}{\omega}\Sigma\frac{1}{x+2\omega}=\frac{2}{\omega}\Sigma\frac{1}{x+\omega}-\frac{2}{\omega}\Sigma\frac{1}{x+2\omega}-\frac{1}{\omega x}.$$

But

$$\Sigma\frac{1}{x+\omega}=\Sigma\frac{1}{x+2\omega}-\frac{1}{x+\omega},$$

so that the desired sum is

$$-\frac{1}{\omega x}-\frac{2}{\omega\,(x+\omega)}=\frac{-3x-\omega}{\omega x\,(x+\omega)}.$$

Example 2. *Find the function whose difference is*

$$\frac{3\omega}{x\,(x+3\omega)}.$$

We let this difference be z. Then

$$z=\frac{1}{x}-\frac{1}{x+3\omega}$$

and

$$\Sigma z=\Sigma\frac{1}{x}-\Sigma\frac{1}{x+3\omega}=\Sigma\frac{1}{x+\omega}-\Sigma\frac{1}{x+3\omega}-\frac{1}{x}$$

$$=\Sigma\frac{1}{x+2\omega}-\Sigma\frac{1}{x+3\omega}-\frac{1}{x}-\frac{1}{x+\omega}=-\frac{1}{x}-\frac{1}{x+\omega}-\frac{1}{x+2\omega},$$

which is the desired sum. Whenever the signs of the sums finally cancel each other, we will be able to find the sum. However, if this mutual annihilation does not occur, it signifies that this sum cannot be found.

2

On the Use of Differences in the Theory of Series

37. It is well known that the nature of series can be very well illustrated from first principles through differences. Indeed, arithmetic progressions, which are ordinarily considered first, have this particular property, that their first differences are equal to each other. From this it follows that their second differences and all higher differences will vanish. There are series whose second differences are constant and for this reason are conveniently called of the *second order*, while arithmetic progressions are said to be of the *first order*. Furthermore, series of the *third order* are those whose third differences are constant; those of the *fourth order* and higher orders are those whose fourth and higher differences are constant.

38. In this division there is an infinite number of kinds of series, but by no means can all series be reduced to one of these. There are innumerably many series whose successive differences never reduce to constants. Besides innumerable others, the geometric progressions never have constant differences of any order. For example, consider

$$1,\ 2,\ 4,\ 8,\ 16,\ 32,\ 64,\ 128,\ \ldots$$

$$1,\ 2,\ 4,\ 8,\ 16,\ 32,\ 64,\ \ldots$$

$$1,\ 2,\ 4,\ 8,\ 16,\ 32,\ \ldots.$$

Since the series of differences of each order is equal to the original series, equality of differences is completely excluded. There are many classes of series, of which only one class is such that its differences of various orders finally reduce to a constant; this chapter will be particularly concerned with that class.

39. Two things are especially important concerning the nature of series: the general term and the sum of the series. The general term is an expression that contains each term of the series and for that reason is a function of the variable x such that when $x = 1$ the first term of the series is obtained, the second when we let $x = 2$, the third when $x = 3$, the fourth when $x = 4$, and so forth. When we know the general term of a series we can find any of the terms, even if the law that relates one term to another is not clear. Thus, for example, for $x = 1000$ we immediately know the thousandth term. In the series

$$1, \quad 6, \quad 15, \quad 28, \quad 45, \quad 66, \quad 91, \quad 120, \quad \ldots$$

the general term is $2x^2 - x$. If $x = 1$, this formula gives the first term, 1; when $x = 2$ we obtain the second term, 6; if we let $x = 3$, the third term 15 appears, and so forth. It is clear that for the 100th term we let $x = 100$, and then $2 \cdot 10000 - 100 = 19900$ is the term.

40. *Indices* or *exponents* in a series are the numbers that indicate which term we are concerned with; thus the index of the first term will be 1, that of the second will be 2, of the third 3, and so forth. Thus the indices of any series are usually written in the following way:

Indices	1	2	3	4	5	6	7	...
Terms	A	B	C	D	E	F	G	...

It is thus immediately clear that G is the seventh term of a given series. From this we see that the general term is nothing else than the term of the series whose index is the indefinite number x. First we will discover how to find the general term of a series whose differences, either first, second, or some other difference is constant. Then we will turn our attention to finding the sum.

41. We begin with the first order, which contains arithmetic progressions, whose first differences are constant. Let a be the first term of the series and let the first term of the series of differences be b, which is equal to all other terms of this series. Hence the series has the form:

Indices	1	2	3	4	5	6	...
Terms	a	$a+b$	$a+2b$	$a+3b$	$a+4b$	$a+5b$	...
Differences	b,	b,	b,	b,	b,	b,	...

From this it is immediately clear that the term whose index is x will be $a + (x-1)\,b$ and the general term will be $bx + a - b$. This is formed from both terms of the series itself and terms of the series of differences. If we call the second term of the series a^{I}, since $b = a^{\mathrm{I}} - a$, then the general term is

$$\left(a^{\mathrm{I}} - a\right) x + 2a - a^{\mathrm{I}} = a^{\mathrm{I}}\left(x - 1\right) - a\left(x - 2\right).$$

Hence from our knowledge of the first and second terms of an arithmetic progression we form the general term.

42. Let a be the first term of a series of the second order, let b be the first term of the series of first differences, and let c be the first term of the series of second differences. Then the series with its differences have the following form:

Indices:

$$1, \quad 2, \quad 3, \quad 4, \quad 5, \quad 6, \quad 7$$

Terms:

$$a, \; a+b, \; a+2b+c, \; a+3b+3c, \; a+4b+6c, \; a+5b+10c, \; a+6b+15c$$

First Differences:

$$b, \quad b+c, \quad b+2c, \quad b+3c, \quad b+4c, \quad b+5c, \quad \ldots$$

Second Differences:

$$c, \quad c, \quad c, \quad c, \quad c, \quad \ldots$$

By inspection we conclude that the term with index x will be

$$a + (x-1)\,b + \frac{(x-1)\,(x-2)}{1 \cdot 2} c,$$

and this is the general term of the given series. However, if we let the second term of the series be a^{I} and the third term be a^{II}, since $b = a^{\mathrm{I}} - a$ and $c = a^{\mathrm{II}} - 2a^{\mathrm{I}} + a$, as we understand from the definition of differences (paragraph 10), we have the general term

$$a + (x-1)\left(a^{\mathrm{I}} - a\right) + \frac{(x-1)\,(x-2)}{1 \cdot 2}\left(a^{\mathrm{II}} - 2a^{\mathrm{I}} + a\right).$$

But this reduces to the form

$$\frac{a^{\mathrm{II}}\,(x-1)\,(x-2)}{1 \cdot 2} - \frac{2a^{\mathrm{I}}\,(x-1)\,(x-3)}{1 \cdot 2} + \frac{a\,(x-2)\,(x-3)}{1 \cdot 2},$$

or

$$\frac{a^{\mathrm{II}}}{2}\,(x-1)\,(x-2) - \frac{2a^{\mathrm{I}}}{2}\,(x-1)\,(x-3) + \frac{a}{2}\,(x-2)\,(x-3),$$

or finally,

$$\frac{1}{2}(x-1)(x-2)(x-3)\left(\frac{a^{\mathrm{II}}}{x-3}-\frac{2a^{\mathrm{I}}}{x-2}+\frac{a}{x-1}\right).$$

It follows that the general term is defined by the first three terms of the series.

43. Let a, a^{I}, a^{II}, a^{III}, $a^{\mathrm{IV}},\ldots$ be the terms of a series of the third order, let b, b^{I}, b^{II}, $b^{\mathrm{III}},\ldots$ be its first differences, and let c, c^{I}, c^{II}, $c^{\mathrm{III}},\ldots$ be its second differences, while d, d, $d,\ldots$ are its third differences, which of course are constants:

Indices: 1, 2, 3, 4, 5, 6, ...

Terms: a, a^{I}, a^{II}, a^{III}, a^{IV}, a^{V}, ...

First Differences: b, b^{I}, b^{II}, b^{III}, b^{IV}, ...

Second Differences: c, c^{I}, c^{II}, c^{III}, ...

Third Differences: d, d, d, ...

Since $a^{\mathrm{I}}=a+b$, $a^{\mathrm{II}}=a+2b+c$, $a^{\mathrm{III}}=a+3b+3c+d$, $a^{\mathrm{IV}}=a+4b+6c+4d,\ldots$, the general term, or the term whose index is x, is

$$a+\frac{(x-1)}{1}b+\frac{(x-1)(x-2)}{1\cdot 2}c+\frac{(x-1)(x-2)(x-3)}{1\cdot 2\cdot 3}d,$$

so that the general term is formed from the differences. Since we have

$$b=a^{\mathrm{I}}-a,\qquad c=a^{\mathrm{II}}-2a^{\mathrm{I}}+a,\qquad d=a^{\mathrm{III}}-3a^{\mathrm{II}}+3a^{\mathrm{I}}-a,$$

when these values are substituted, the general term will be

$$a^{\mathrm{III}}\frac{(x-1)(x-2)(x-3)}{1\cdot 2\cdot 3}-3a^{\mathrm{II}}\frac{(x-1)(x-2)(x-4)}{1\cdot 2\cdot 3}$$
$$+3a^{\mathrm{I}}\frac{(x-1)(x-3)(x-4)}{1\cdot 2\cdot 3}-a\frac{(x-2)(x-3)(x-4)}{1\cdot 2\cdot 3}.$$

This can also be expressed as

$$\frac{(x-1)(x-2)(x-3)(x-4)}{1\cdot 2\cdot 3}\left(\frac{a^{\mathrm{III}}}{x-4}-\frac{3a^{\mathrm{II}}}{x-3}+\frac{3a^{\mathrm{I}}}{x-2}-\frac{a}{x-1}\right).$$

44. Now let a series of any order be given:

Indices: 1, 2, 3, 4, 5, 6, ...

Terms: $a, \quad a^{\mathrm{I}}, \quad a^{\mathrm{II}}, \quad a^{\mathrm{III}}, \quad a^{\mathrm{IV}}, \quad a^{\mathrm{V}}, \quad \ldots$

First Differences: $b, \quad b^{\mathrm{I}}, \quad b^{\mathrm{II}}, \quad b^{\mathrm{III}}, \quad b^{\mathrm{IV}}, \quad \ldots$

Second Differences: $c, \quad c^{\mathrm{I}}, \quad c^{\mathrm{II}}, \quad c^{\mathrm{III}}, \quad \ldots$

Third Differences: $d, \quad d^{\mathrm{I}}, \quad d^{\mathrm{II}}, \quad \ldots$

Fourth Differences: $e, \quad e^{\mathrm{I}}, \quad \ldots$

Fifth Differences: $f, \quad \ldots$

and so forth. From the first term of the series and the first terms of the differences, b, c, d, e, $f, \ldots$ we can express the general term as

$$a + \frac{(x-1)}{1} b + \frac{(x-1)(x-2)}{1 \cdot 2} c + \frac{(x-1)(x-2)(x-3)}{1 \cdot 2 \cdot 3} d + \frac{(x-1)(x-2)(x-3)(x-4)}{1 \cdot 2 \cdot 3 \cdot 4} e + \cdots$$

until we come to the constant differences. From this it is clear that if we never produce constant differences, then the general term will be expressed by an infinite series.

45. Since the differences are formed from the terms of the given series, if these values are substituted, the general term in this form for any series of the first, second, and third orders have been given. For a series of the fourth order the general term is

$$\frac{(x-1)(x-2)(x-3)(x-4)(x-5)}{1 \cdot 2 \cdot 3 \cdot 4} \times \left(\frac{a^{\mathrm{IV}}}{x-5} - \frac{4a^{\mathrm{III}}}{x-4} + \frac{6a^{\mathrm{II}}}{x-3} - \frac{4a^{\mathrm{I}}}{x-2} + \frac{a}{x-1} \right).$$

From this the law of formation for the general term for higher-order sequences is easily seen. It is also clear that for any order the general term will be a polynomial in x whose degree will be no higher than the order of the series to which it refers. Thus, a series of the first order has a general term that is a first-degree function, a second-order series has a second-degree term, and so forth.

46. The differences, as we have seen above, can be expressed in terms of the original series as follows:

$$
\begin{array}{lll}
b = a^{\mathrm{I}} - a, & c = a^{\mathrm{II}} - 2a^{\mathrm{I}} + a, & d = a^{\mathrm{III}} - 3a^{\mathrm{II}} + 3a^{\mathrm{I}} - a, \\
b^{\mathrm{I}} = a^{\mathrm{II}} - a^{\mathrm{I}}, & c^{\mathrm{I}} = a^{\mathrm{III}} - 2a^{\mathrm{II}} + a^{\mathrm{I}}, & d^{\mathrm{I}} = a^{\mathrm{IV}} - 3a^{\mathrm{III}} + 3a^{\mathrm{II}} - a^{\mathrm{I}}, \\
b^{\mathrm{II}} = a^{\mathrm{III}} - a^{\mathrm{II}}, & c^{\mathrm{II}} = a^{\mathrm{IV}} - 2a^{\mathrm{III}} + a^{\mathrm{II}}, & d^{\mathrm{II}} = a^{\mathrm{V}} - 3a^{\mathrm{IV}} + 3a^{\mathrm{III}} - a^{\mathrm{II}}, \\
\ldots, & \ldots, & \ldots.
\end{array}
$$

Since in a series of the first order all values of c vanish, we have

$$a^{\mathrm{II}} = 2a^{\mathrm{I}} - a, \qquad a^{\mathrm{III}} = 2a^{\mathrm{II}} - a^{\mathrm{I}}, \qquad a^{\mathrm{IV}} = 2a^{\mathrm{III}} - a^{\mathrm{II}}, \qquad \ldots,$$

and so it is clear that these series are recurrent and that the scale of relation is $2, -1$. Then, since in a series of the second order all values of d vanish, we have

$$a^{\mathrm{III}} = 3a^{\mathrm{II}} - 3a^{\mathrm{I}} + a, \qquad a^{\mathrm{IV}} = 3a^{\mathrm{III}} - 3a^{\mathrm{II}} + a^{\mathrm{I}}, \qquad \ldots.$$

From this it follows that these series are recurrent with a scale of relation $3, -3, 1$. In a similar way it can be shown that each such series of any order is both a recurrent series and the scale of relation consists of the binomial coefficients where the exponent is one more than the order of the series.

47. Since in a series of the first order we also have all d's, e's, and all of the subsequent differences vanishing, we also have

$$
\begin{aligned}
a^{\mathrm{III}} &= 3a^{\mathrm{II}} - 3a^{\mathrm{I}} + a, \\
a^{\mathrm{IV}} &= 3a^{\mathrm{III}} - 3a^{\mathrm{II}} + a^{\mathrm{I}},
\end{aligned}
$$

and so forth, or

$$
\begin{aligned}
a^{\mathrm{IV}} &= 4a^{\mathrm{III}} - 6a^{\mathrm{II}} + 4a^{\mathrm{I}} - a, \\
a^{\mathrm{V}} &= 4a^{\mathrm{IV}} - 6a^{\mathrm{III}} + 4a^{\mathrm{II}} - a^{\mathrm{I}},
\end{aligned}
$$

and so forth. From this it follows that these are recurrent series, indeed in an infinite number of ways, since the scales of relation can be $3, -3, +1$; $4, -6, +4, -1$; $5, -10, +10, -5 + 1$; .... In a similar way it should be understood that each of the series of the kind we have been discussing is a recurrent series in an infinite number of ways and that the scale of relation is

$$n, \quad -\frac{n(n-1)}{1 \cdot 2}, \quad +\frac{n(n-1)(n-2)}{1 \cdot 2 \cdot 3}, \quad -\frac{n(n-1)(n-2)(n-3)}{1 \cdot 2 \cdot 3 \cdot 4},$$

and so forth, provided that n is an integer and is larger than the order of the given series. These series can also arise from fractions whose denominator is $(1-y)^n$, as is shown in a previous book[1] where recurrent series are treated at greater length.

48. As we have seen, every series of this class, no matter of what order, has a general term that is a polynomial. On the other hand, we will see that every series that has this kind of function for its general term belongs to this class of series and the differences eventually are a constant. Indeed, if the general term is a second-degree polynomial of the form ax^2+bx+c, then the series obtained by letting x be equal successively to $1, 2, 3, 4, 5, \ldots$, will be of the second order, and the second difference will be constant. Likewise, if the general term is a third-degree polynomial ax^3+bx^2+cx+d, then the series will be of the third order, and so forth.

49. From the general term we can find not only all of the terms of the series, but also the series of differences, both the first differences and also the higher differences. If the first term of a series is subtracted from the second, we obtain the first term of the series of differences. Likewise, we obtain the second term of this series if we subtract the second term from the third term of the original series. Thus we obtain the term of the series of differences whose index is x if we subtract the term of the original series whose index is x from the term whose index is $x+1$. Hence, if in the general term of the series we substitute $x+1$ for x and from this subtract the general term, the remainder will be the general term of the series of differences. If X is the general term of the series, then its difference ΔX (which is obtained in the way shown in the previous chapter if we let $\omega = 1$) is the general term of the series of first differences. Likewise, $\Delta^2 X$ is the general term of the series of second differences; $\Delta^3 X$ is the general term of the series of third differences, and so forth.

50. If the general term X is a polynomial of degree n, from the previous chapter we know that its difference ΔX is a polynomial of degree $n-1$. Furthermore, $\Delta^2 X$ is of degree $n-2$, and $\Delta^3 X$ is of degree $n-3$, and so forth. It follows that if X is a first-degree polynomial, as is $ax+b$, then its difference ΔX is the constant a. Since this is the general term of the series of first differences, we see that a series whose general term is a first-degree polynomial is an arithmetic progression, that is, a series of the first order. Likewise, if the general term is a second-degree polynomial, since $\Delta^2 X$ is a constant, the series of second differences is constant and the original series is of second order. In like manner it is always true that the degree of the general term is the order of the series it defines.

[1] *Introduction*, Book I, Chapter IV.

51. It follows that a series of powers of the natural numbers will have constant differences, as is clear from the following scheme:

First Powers: 1, 2, 3, 4, 5, 6, 7, 8, ...

First Differences: 1, 1, 1, 1, 1, 1, 1, ...

Second Powers: 1, 4, 9, 36, 49, 64, ...

First Differences: 3, 5, 7, 9, 11, 13, 15, ...

Second Differences: 2, 2, 2, 2, 2, 2, ...

Third Powers: 1, 8, 27, 64, 125, 216, 343, ...

First Differences: 7, 19, 37, 61, 91, 127, ...

Second Differences: 12, 18, 24, 30, 36, ...

Third Differences: 6, 6, 6, 6, ...

Fourth Powers: 1, 16, 81, 256, 625, 1296, 2401, ...

First Differences: 15, 65, 175, 369, 671, 1105, ...

Second Differences: 50, 110, 194, 302, 434, ...

Third Differences: 60, 84, 108, 132, ...

Fourth Differences: 24, 24, 24, ...

The rules given in the previous chapter for finding differences of any order can now be used to find the general terms for differences of any order for a given series.

52. If the general term for any series is known, then not only can it be used to find all of the terms, but it can also be used to reverse the order and find terms with negative indices, by substituting negative values for x. For example, if the general term is $\left(x^2+3x\right)/2$ and we use both negative and positive indices, we can continue the series in both ways as follows:

Indices: $\ldots, -5, -4, -3, -2, -1, 0, 1, 2, 3, 4, 5, 6, \ldots$

Series: $\ldots, 5, 2, 0, -1, -1, 0, 2, 5, 9, 14, 20, 27, \ldots$

First Differences: $\ldots, -3, -2, -1, 0, 1, 2, 3, 4, 5, 6, 7, \ldots$

Second Differences: $\ldots, 1, 1, 1, 1, 1, 1, 1, 1, 1, 1, \ldots$

Since the general term is formed from the differences, any series can be continued backwards so that if the differences finally are constant, the general term can be expressed in finite form. If the differences are not finally constant, then the general term requires an infinite expression. From the general term we can also define terms whose indices are fractions, and this gives an *interpolation* of the series.

53. After these remarks about the general terms of series we now turn to the investigation of the sum, or general partial sum, of a series of any order. Given any series the *general partial sum* is a function of x that is equal to the sum of x terms of the series. Hence the general partial sum will be such that if $x = 1$, then it will be equal to the first term of the series. If $x = 2$, then it gives the sum of the first two terms of the series; let $x = 3$, and we have the sum of the first three terms, and so forth. Therefore, if from a given series we form a new series whose first term is equal to the first term of the given series, second term is the sum of the first two terms of the given series, the third of the first three terms, and so forth, then this new series is its *partial sum series.* The general term of this new series is the general partial sum. Hence, finding the general partial sum brings us back to finding the general term of a series.

54. Let the given series be

$$a, \quad a^{\mathrm{I}}, \quad a^{\mathrm{II}}, \quad a^{\mathrm{III}}, \quad a^{\mathrm{IV}}, \quad a^{\mathrm{V}}, \quad \ldots$$

and let the series of partial sums be

$$A, \quad A^{\mathrm{I}}, \quad A^{\mathrm{II}}, \quad A^{\mathrm{III}}, \quad A^{\mathrm{IV}}, \quad A^{\mathrm{V}}, \quad \ldots.$$

From the definition we have

$$\begin{aligned} A &= a, \\ A^{\mathrm{I}} &= a + a^{\mathrm{I}}, \\ A^{\mathrm{II}} &= a + a^{\mathrm{I}} + a^{\mathrm{II}}, \\ A^{\mathrm{III}} &= a + a^{\mathrm{I}} + a^{\mathrm{II}} + a^{\mathrm{III}}, \\ A^{\mathrm{IV}} &= a + a^{\mathrm{I}} + a^{\mathrm{II}} + a^{\mathrm{III}} + a^{\mathrm{IV}}, \\ A^{\mathrm{V}} &= a + a^{\mathrm{I}} + a^{\mathrm{II}} + a^{\mathrm{III}} + a^{\mathrm{IV}} + a^{\mathrm{V}}, \end{aligned}$$

and so forth. Now, the series of differences of the series of partial sums is

$$A^{\mathrm{I}} - A = a^{\mathrm{I}}, \qquad A^{\mathrm{II}} - A^{\mathrm{I}} = a^{\mathrm{II}}, \qquad A^{\mathrm{III}} - A^{\mathrm{II}} = a^{\mathrm{III}}, \qquad \ldots,$$

so that if we remove the first term of the given series, we have the series of first differences of the series of partial sums. If we supply a zero as the first term of the series of partial sums, to give

$$0, \quad A, \quad A^{\mathrm{I}}, \quad A^{\mathrm{II}}, \quad A^{\mathrm{III}}, \quad A^{\mathrm{IV}}, \quad A^{\mathrm{V}}, \quad \ldots,$$

then the series of first differences is the given series

$$a, \quad a^{\mathrm{I}}, \quad a^{\mathrm{II}}, \quad a^{\mathrm{III}}, \quad a^{\mathrm{IV}}, \quad a^{\mathrm{V}}, \quad \ldots.$$

55. For this reason, the first differences of the given series are the second differences of the series of partial sums; the second differences of the former are the third differences of the latter; the third of the former are the fourth of the latter, and so forth. Hence, if the given series finally has constant differences, then the series of partial sums also eventually has constant differences and so is of the same kind except one order higher. It follows that this kind of series always has a partial sum that can be given as a finite expression. Indeed, the general term of the series

$$0, \quad A, \quad A^{\mathrm{I}}, \quad A^{\mathrm{II}}, \quad A^{\mathrm{III}}, \quad A^{\mathrm{IV}}, \quad \ldots,$$

or that expression which corresponds to x, gives the sum of the $x-1$ terms of the series a, a^{I}, a^{II}, a^{III}, a^{IV}, $\ldots$. If instead of x we write $x+1$, we obtain the sum of x terms which is the general term.

56. Let a given series be

$$a, \quad a^{\mathrm{I}}, \quad a^{\mathrm{II}}, \quad a^{\mathrm{III}}, \quad a^{\mathrm{IV}}, \quad a^{\mathrm{V}}, \quad a^{\mathrm{VI}}, \quad \ldots.$$

The series of first differences is

$$b, \quad b^{\mathrm{I}}, \quad b^{\mathrm{II}}, \quad b^{\mathrm{III}}, \quad b^{\mathrm{IV}}, \quad b^{\mathrm{V}}, \quad b^{\mathrm{VI}}, \quad \ldots,$$

the series of second differences is

$$c, \quad c^{\mathrm{I}}, \quad c^{\mathrm{II}}, \quad c^{\mathrm{III}}, \quad c^{\mathrm{IV}}, \quad c^{\mathrm{V}}, \quad c^{\mathrm{VI}}, \quad \ldots,$$

the series of third differences is

$$d, \quad d^{\mathrm{I}}, \quad d^{\mathrm{II}}, \quad d^{\mathrm{III}}, \quad d^{\mathrm{IV}}, \quad d^{\mathrm{V}}, \quad d^{\mathrm{VI}}, \quad \ldots,$$

and so forth, until we come to constant differences. We then form the series of partial sums, with 0 as its first term, and the succeeding differences in the following way:

Indices: $1, \quad 2, \quad 3, \quad 4, \quad 5, \quad 6, \quad 7, \quad \ldots$

Partial Sums: $0, \quad A, \quad A^{\mathrm{I}}, \quad A^{\mathrm{II}}, \quad A^{\mathrm{III}}, \quad A^{\mathrm{IV}}, \quad A^{\mathrm{V}}, \quad \ldots$

Given Series: $a, \quad a^{\mathrm{I}}, \quad a^{\mathrm{II}}, \quad a^{\mathrm{III}}, \quad a^{\mathrm{IV}}, \quad a^{\mathrm{V}}, \quad a^{\mathrm{VI}}, \quad \ldots$

First Differences: $b, \quad b^{\mathrm{I}}, \quad b^{\mathrm{II}}, \quad b^{\mathrm{III}}, \quad b^{\mathrm{IV}}, \quad b^{\mathrm{V}}, \quad b^{\mathrm{VI}}, \quad \ldots$

Second Differences: $c, \quad c^{\mathrm{I}}, \quad c^{\mathrm{II}}, \quad c^{\mathrm{III}}, \quad c^{\mathrm{IV}}, \quad c^{\mathrm{V}}, \quad c^{\mathrm{VI}}, \quad \ldots$

Third Differences: $d, \quad d^{\mathrm{I}}, \quad d^{\mathrm{II}}, \quad d^{\mathrm{III}}, \quad d^{\mathrm{IV}}, \quad d^{\mathrm{V}}, \quad d^{\mathrm{VI}}, \quad \ldots$

The general term of the series of partial sums, that is, the term corresponding to the index x, is

$$0 + (x-1)\,a + \frac{(x-1)\,(x-2)}{1\cdot 2}b + \frac{(x-1)\,(x-2)\,(x-3)}{1\cdot 2\cdot 3}c + \cdots,$$

but this is also the series of partial sums of the first $x-1$ terms of the given series a, a^{I}, a^{II}, $a^{\mathrm{IV}}, \ldots$.

57. Hence, if instead of $x-1$ we write x, we obtain the series of partial sums

$$xa + \frac{x\,(x-1)}{1\cdot 2}b + \frac{x\,(x-1)\,(x-2)}{1\cdot 2\cdot 3}c + \frac{x\,(x-1)\,(x-2)\,(x-3)}{1\cdot 2\cdot 3\cdot 4}d + \cdots.$$

If we let the letters b, c, d, $e, \ldots$ keep the assigned values, then we have

Series: $a, \quad a^{\mathrm{I}}, \quad a^{\mathrm{II}}, \quad a^{\mathrm{III}}, \quad a^{\mathrm{IV}}, \quad a^{\mathrm{V}}, \quad \ldots$

General Term:

$$a + (x-1)\,b + \frac{(x-1)\,(x-2)}{1\cdot 2}c + \frac{(x-1)\,(x-2)\,(x-3)}{1\cdot 2\cdot 3}d$$
$$+ \frac{(x-1)\,(x-2)\,(x-3)\,(x-4)}{1\cdot 2\cdot 3\cdot 4}e + \cdots,$$

Partial Sum:

$$xa + \frac{x\,(x-1)}{1\cdot 2}b + \frac{x\,(x-1)\,(x-2)}{1\cdot 2\cdot 3}c + \frac{x\,(x-1)\,(x-2)\,(x-3)}{1\cdot 2\cdot 3\cdot 4}d + \cdots.$$

Therefore, once a series of any order is found, the general term can easily be found from the partial sum in the way we have shown, namely, by combining differences.

58. This method of finding the partial sum of a series through differences is most useful for those series whose differences eventually become constant. In other cases we do not obtain a finite expression. If we pay close attention to the character of the partial sums, which we have already discussed, then another method is open to us for finding the partial sum immediately from the general term. Indeed this method is much more general, and in the infinite case we obtain finite expressions, rather than the infinite that we obtain from the previous method. Let a given series be

$$a, \qquad b, \qquad c, \qquad d, \qquad e, \qquad f, \qquad \ldots,$$

with X the general term corresponding to the index x, and S the partial sum. Since S is the sum of the first x terms, the sum of the first $x-1$ terms is $S-X$. Furthermore, X is the difference, since this is what remains when $S-X$ is subtracted from the next term S.

59. Since $X = \Delta(S-X)$ is the difference, as defined in the previous chapter, provided that we let the constant ω equal 1, if we return to the sum, then we have $\Sigma X = S - X$, and the desired partial sum is

$$S = \Sigma X + X + C.$$

Hence we need to find the sum of the function X by the method previously discussed, to which we add the general term X to obtain the partial sum. Since this process involves a constant quantity that must either be added or subtracted, we need to accommodate this to the present case. It is clear that if we let $x = 0$, the number of terms in the sum is zero, and the sum also should be zero. From this fact the constant C should be calculated by letting both $x = 0$ and $S = 0$. In the expression $S = \Sigma X + X + C$ with both $S = 0$ and $x = 0$, we obtain the value of C.

60. Since this whole business reduces to the sums of functions we found (in paragraph 27) when $\omega = 1$, we recall those results, especially for the powers of x:

$$\begin{aligned}
\Sigma x^0 &= \Sigma 1 = x, \\
\Sigma x &= \frac{1}{2}x^2 - \frac{1}{2}x, \\
\Sigma x^2 &= \frac{1}{3}x^3 - \frac{1}{2}x^2 + \frac{1}{6}x, \\
\Sigma x^3 &= \frac{1}{4}x^4 - \frac{1}{2}x^3 + \frac{1}{4}x^2, \\
\Sigma x^4 &= \frac{1}{5}x^5 - \frac{1}{2}x^4 + \frac{1}{3}x^3 - \frac{1}{30}x, \\
\Sigma x^5 &= \frac{1}{6}x^6 - \frac{1}{2}x^5 + \frac{5}{12}x^4 - \frac{1}{12}x^2, \\
\Sigma x^6 &= \frac{1}{7}x^7 - \frac{1}{2}x^6 + \frac{1}{2}x^5 - \frac{1}{6}x^3 + \frac{1}{42}x.
\end{aligned}$$

In paragraph 29 we gave the sum for a general power of x, provided that everywhere we let $\omega = 1$. With these formulas we can easily find the partial sum, provided that the general term is a polynomial in x.

61. Let $S.X$ be the partial sum of the series whose general term is X. As we have seen,

$$S.X = \Sigma X + X + C,$$

where C is the constant we obtain by letting $S.X$ vanish when we let $x = 0$. It follows that we can express the partial sum of the series of powers, that is, the series for which the general term has the form x^n. We let

$$S.X = 1 + 2^n + 3^n + 4^n + \cdots + x^n.$$

Then

$$\begin{aligned} S.x^n = {} & \frac{1}{n+1}x^{n+1} + \frac{1}{2}x^n + \frac{1}{2}\cdot\frac{n}{2\cdot 3}x^{n-1} - \frac{1}{6}\cdot\frac{n(n-1)(n-2)}{2\cdot 3\cdot 4\cdot 5}x^{n-3} \\ & + \frac{1}{6}\cdot\frac{n(n-1)(n-2)(n-3)(n-4)}{2\cdot 3\cdots 6\cdot 7}x^{n-5} \\ & - \frac{3}{10}\cdot\frac{n(n-1)\cdots(n-6)}{2\cdot 3\cdots 8\cdot 9}x^{n-7} \\ & + \frac{5}{6}\cdot\frac{n(n-1)\cdots(n-8)}{2\cdot 3\cdots 10\cdot 11}x^{n-9} - \frac{691}{210}\cdot\frac{n(n-1)\cdots(n-10)}{2\cdot 3\cdots 12\cdot 13}x^{n-11} \\ & + \frac{35}{2}\cdot\frac{n(n-1)\cdots(n-12)}{2\cdot 3\cdots 14\cdot 15}x^{n-13} \\ & - \frac{3617}{30}\cdot\frac{n(n-1)\cdots(n-14)}{2\cdot 3\cdots 16\cdot 17}x^{n-15} \\ & + \frac{43867}{42}\cdot\frac{n(n-1)\cdots(n-16)}{2\cdot 3\cdots 18\cdot 19}x^{n-17} \\ & - \frac{1222277}{110}\cdot\frac{n(n-1)\cdots(n-18)}{2\cdot 3\cdots 20\cdot 21}x^{n-19} \\ & + \frac{854513}{6}\cdot\frac{n(n-1)\cdots(n-20)}{2\cdot 3\cdots 22\cdot 23}x^{n-21} \\ & - \frac{1181820455}{546}\cdot\frac{n(n-1)\cdots(n-22)}{2\cdot 3\cdots 24\cdot 25}x^{n-23} \\ & + \frac{76977927}{2}\cdot\frac{n(n-1)\cdots(n-24)}{2\cdot 3\cdots 26\cdot 27}x^{n-25} \\ & - \frac{23749461029}{30}\cdot\frac{n(n-1)\cdots(n-26)}{2\cdot 3\cdots 28\cdot 29}x^{n-27}, \end{aligned}$$

and so forth.

62. Hence, for various values of n we have the following:

$$S.x^0 = x,$$

$$S.x^1 = \frac{1}{2}x^2 + \frac{1}{2}x,$$

$$S.x^2 = \frac{1}{3}x^3 + \frac{1}{2}x^2 + \frac{1}{6}x,$$

$$S.x^3 = \frac{1}{4}x^4 + \frac{1}{2}x^3 + \frac{1}{4}x^2,$$

$$Sx^4 = \frac{1}{5}x^5 + \frac{1}{2}x^4 + \frac{1}{3}x^3 - \frac{1}{30}x,$$

$$S.x^5 = \frac{1}{6}x^6 + \frac{1}{2}x^5 + \frac{5}{12}x^4 - \frac{1}{12}x^2,$$

$$S.x^6 = \frac{1}{7}x^7 + \frac{1}{2}x^6 + \frac{1}{2}x^5 - \frac{1}{6}x^3 + \frac{1}{42}x,$$

$$S.x^7 = \frac{1}{8}x^8 + \frac{1}{2}x^7 + \frac{7}{12}x^6 - \frac{7}{24}x^4 + \frac{1}{12}x^2,$$

$$S.x^8 = \frac{1}{9} + \frac{1}{2}x^8 + \frac{2}{3}x^7 - \frac{7}{15}s^5 + \frac{2}{9}x^3 - \frac{1}{30}x,$$

$$S.x^9 = \frac{1}{10}x^{10} + \frac{1}{2}x^9 + \frac{3}{4}x^8 - \frac{7}{10}x^6 + \frac{1}{2}x^4 - \frac{3}{20}x^2,$$

$$S.x^{10} = \frac{1}{11}x^{11} + \frac{1}{2}x^{10} + \frac{5}{6}x^9 - x^7 + x^5 - \frac{1}{2}x^3 + \frac{5}{66}x,$$

$$S.x^{11} = \frac{1}{12}x^{12} + \frac{1}{2}x^{11} + \frac{11}{12}x^{10} - \frac{11}{8}x^8 + \frac{11}{6}x^6 - \frac{11}{8}x^4 + \frac{5}{12}x^2,$$

$$S.x^{12} = \frac{1}{13}x^{13} + \frac{1}{2}x^{12} + x^{11} - \frac{11}{6}x^9 + \frac{22}{7}x^7 - \frac{33}{10}x^5 + \frac{5}{3}x^3 - \frac{691}{2730}x,$$

$$S.x^{13} = \frac{1}{14}x^{14} + \frac{1}{2}x^{13} + \frac{13}{12}x^{12} - \frac{143}{60}x^{10} + \frac{143}{28}x^8 - \frac{143}{20}x^6 - \frac{691}{420}x^2,$$

$$S.x^{14} = \frac{1}{15}x^{15} + \frac{1}{2}x^{14} + \frac{7}{6}x^{13} - \frac{91}{30}x^{11} + \frac{143}{18}x^9 - \frac{143}{10}x^7$$
$$+ \frac{91}{6}x^5 - \frac{691}{90}x^3 + \frac{7}{6}x,$$

$$S.x^{15} = \frac{1}{16}x^{16} + \frac{1}{2}x^{15} + \frac{5}{4}x^{14} - \frac{91}{24}x^{12} + \frac{143}{12}x^{10} - \frac{429}{16}x^8$$
$$+ \frac{455}{12}x^6 - \frac{691}{24}x^4 + \frac{35}{4}x^2,$$

$$S.x^{16} = \frac{1}{17}x^{17} + \frac{1}{2}x^{16} + \frac{4}{3}x^{15} - \frac{14}{3}x^{13} + \frac{52}{3}x^{11} - \frac{143}{3}x^9$$
$$+ \frac{260}{3}x^7 - \frac{1382}{15}x^5 + \frac{140}{3}x^3 - \frac{3617}{510}x,$$

and so forth. These formulas can be continued to the twenty-ninth power. Indeed, they can be carried to even higher powers, provided that the numerical coefficients have been worked out.

63. In these formulas we can observe a general law, with whose help we can easily find any one of the formulas from the preceding formula, except for the last term, in case the power of that term is 1. In that case we have to find one more term. If we ignore that term for the moment, and if

$$S.x^n = \alpha x^{n+1} + \beta x^n + \gamma x^{n-1} - \delta x^{n-3} + \epsilon x^{n-5} - \zeta x^{n-7} + \eta x^{n-9} - \cdots,$$

then the subsequent formula will be

$$S.x^{x+1} = \frac{n+1}{n+2}\alpha x^{n+2} - \frac{n+1}{n+1}\beta x^{n+1} + \frac{n+1}{n}\gamma x^n - \frac{n+1}{n-2}\delta x^{n-2}$$
$$+ \frac{n+1}{n-4}\epsilon x^{n-4} - \frac{n+1}{n-6}\zeta x^{n-6} + \frac{n+1}{n-8}\eta x^{n-8} - \cdots,$$

and if n is even, we obtain the true formula. If n is odd, then the formula is lacking a term, which has the form $\pm\phi x$. Now with very little work we can discover ϕ. Since when we let $x = 1$ the partial sum should be but a single term, the first term, and this is equal to 1, if we let $x = 1$ in all of the terms found so far, the sum should be equal to 1. In this way we evaluate ϕ. Once this is found we can proceed to the next step. In this way all of these sums can be found. Thus, since

$$S.x^5 = \frac{1}{6}x^6 + \frac{1}{2}x^5 + \frac{5}{12}x^4 - \frac{1}{12}x^2,$$

we have

$$S.x^6 = \frac{6}{7}\cdot\frac{1}{6}x^7 + \frac{6}{6}\cdot\frac{1}{2}x^6 + \frac{6}{5}\cdot\frac{5}{12}x^5 - \frac{6}{3}\cdot\frac{1}{12}x^3 + \phi x,$$

or

$$S.x^6 = \frac{1}{7}x^7 + \frac{1}{2}x^6 + \frac{1}{2}x^5 - \frac{1}{6}x^3 + \phi x.$$

Now let $x = 1$, so that $1 = \frac{1}{7} + \frac{1}{2} + \frac{1}{2} - \frac{1}{6} + \phi$, and so $\phi = \frac{1}{6} - \frac{1}{7} = \frac{1}{42}$, just as we found in the general form.

64. By means of these formulas for partial sums we can easily find the partial sums of all series whose general term is a polynomial, and much more expeditiously than by the previous method using differences.

Example 1. *Find the partial sum of the series* $2, 7, 15, 26, 40, 57, 77, 100, 126, \ldots,$ *whose general term is* $(3x^2 + x)/2$.

Since the general term consists of two members, we find the partial sum for each of them from the above formulas

$$S.\frac{3}{2}x^2 = \frac{1}{2}x^3 + \frac{3}{4}x^2 + \frac{1}{4}x$$

and

$$S.\frac{1}{2}x = \frac{1}{4}x^2 + \frac{1}{4}x,$$

so that

$$S.\frac{3x^2+x}{2} = \frac{1}{2}x^3 + x^2 + \frac{1}{2}x = \frac{1}{2}x\left(x+1\right)^2,$$

and this is the desired partial sum. Thus if $x = 5$, we have $\frac{5}{2} \cdot 6^2 = 90$, while the sum of the terms is

$$2 + 7 + 15 + 26 + 40 = 90.$$

Example 2. *Find the partial sum of the series* $1, 27, 125, 343, 729, 1331$, $\ldots$, *which is the sum of the cubes of the odd integers.*

The general term of this series is

$$(2x-1)^3 = 8x^3 - 12x^2 + 6x - 1,$$

so we collect the partial sums in the following way:

$$+8.S.x^3 = 2x^4 + 4x^3 + 2x^2,$$

$$-12.S.x^2 = -4x^3 - 6x^2 - 2x,$$

$$+6.S.x = +3x^2 + 3x,$$

and

$$-1.S.x^0 = -x.$$

Then the desired sum is

$$2x^4 - x^2 = x^2\left(2x^2 - 1\right).$$

Hence, if $x = 6$, we have $36 \cdot 71 = 2556$, which is the sum of the first six terms of the given series:

$$1 + 27 + 125 + 343 + 729 + 1331 = 2556.$$

65. If the general term is a product of linear factors as in paragraph 32, then it is easier to find the partial sums by the method treated in that section and the sections following. When we let $\omega = 1$, we have

$$\Sigma\,(x+n) = \frac{1}{2}\,(x+n-1)\,(x+n)\,,$$

$$\Sigma\,(x+n)\,(x+n+1) = \frac{1}{3}\,(x+n-1)\,(x+n)\,(x+n+1)\,,$$

$$\Sigma\,(x+n)\,(x+n+1)\,(x+n+2) = \frac{1}{4}\,(x+n-1)\,(x+n)\,(x+n+1) \times (x+n+2)\,,$$

and so forth. If we add to these sums the general term and a constant, which can be calculated by letting $x = 0$, with the partial sum vanishing, then we obtain the following:

$$S.\,(x+n) = \frac{1}{2}\,(x+n)\,(x+n+1) - \frac{1}{2}n\,(n+1)\,,$$

$$S.\,(x+n)\,(x+n+1) = \frac{1}{3}\,(x+n)\,(x+n+1)\,(x+n+2) - \frac{1}{3}n\,(n+1)\,(n+2)\,,$$

$$S.\,(x+n)\,(x+n+1)\,(x+n+2) = \frac{1}{4}\,(x+n)\,(x+n+1)\,(x+n+2) \times (x+n+3) - \frac{1}{4}n\,(n+1)\,(n+2)\,(n+3)\,,$$

and so forth.

If we let $n = 0$ or $n = -1$, then the constant in these partial sums will vanish.

66. For the series $1, 2, 3, 4, 5, \ldots$, whose general term is x, the partial sum is $\frac{1}{2}x\,(x+1)$. The series of partial sums, $1, 3, 6, 10, 15, \ldots$ has a partial sum

$$\frac{x\,(x+1)\,(x+2)}{1 \cdot 2 \cdot 3}.$$

This series of partial sums $1, 4, 10, 20, 35, \ldots$ has a partial sum

$$\frac{x\,(x+1)\,(x+2)\,(x+3)}{1 \cdot 2 \cdot 3 \cdot 4},$$

which is the general term of the series $1, 5, 15, 35, 70, \ldots$ with partial sum

$$\frac{x\,(x+1)\,(x+2)\,(x+3)\,(x+4)}{1 \cdot 2 \cdot 3 \cdot 4 \cdot 5}.$$

These series have a special importance, since they have many applications. For instance, from these we obtain coefficients of binomials with high degrees, and it is clear to anyone with some experience in this area that these are important.

67. We can also use these to find more easily the partial sums that we formerly found with differences. Since we found the general term to be of the form

$$a + \frac{x-1}{1} b + \frac{(x-1)(x-2)}{1 \cdot 2} c + \frac{(x-1)(x-2)(x-3)}{1 \cdot 2 \cdot 3} d + \cdots ,$$

if we take the partial sum of each member and then add all of these, we will have the partial sum of the series with the given general term. Thus, since

$$S.1 = x,$$

$$S.(x-1) = \frac{1}{2} x (x-1),$$

$$S.(x-1)(x-2) = \frac{1}{3} x (x-1)(x-2),$$

$$S.(x-1)(x-2)(x-3) = \frac{1}{4} x (x-1)(x-2)(x-3),$$

and so forth, we have the desired partial sum

$$xa + \frac{x(x-1)}{1 \cdot 2} b + \frac{x(x-1)(x-2)}{1 \cdot 2 \cdot 3} c + \frac{x(x-1)(x-2)(x-3)}{1 \cdot 2 \cdot 3 \cdot 4} d + \cdots ,$$

and this is exactly the form we obtained before in paragraph 57 with differences.

68. We can also find these partial sums for quotients. Since previously, in paragraph 34, we obtained, when $\omega = 1$,

$$\Sigma \frac{1}{(x+n)(x+n+1)} = -1 \cdot \frac{1}{x+n},$$

so that

$$S. \frac{1}{(x+n)(x+n+1)} = -1 \cdot \frac{1}{x+n+1} + \frac{1}{n+1}.$$

In a similar way, if we add to the above sum the general term, or, what comes to the same thing, if in these expressions we substitute $x+1$ for x, then we have

$$S. \frac{1}{(x+n)(x+n+1)(x+n+2)}$$

$$= -\frac{1}{2} \cdot \frac{1}{(x+n+1)(x+n+2)} + \frac{1}{2} \cdot \frac{1}{(n+1)(n+2)}$$

and

$$S.\frac{1}{(x+n)(x+n+1)(x+n+2)(x+n+3)}$$

$$= -\frac{1}{3}\cdot\frac{1}{(x+n+1)(x+n+2)(x+n+3)}$$

$$+\frac{1}{3}\cdot\frac{1}{(n+1)(n+2)(n+3)}.$$

These forms can easily be continued as far as desired.

69. Since

$$S.\frac{1}{(x+n)(x+n+1)} = \frac{1}{n+1} - \frac{1}{x+n+1},$$

we also have

$$S.\frac{1}{x+n} - S.\frac{1}{x+n+1} = \frac{1}{n+1} - \frac{1}{x+n+1}.$$

Although neither of these two partial sums can be expressed by itself, still their difference is known. In many cases, by this means the sum of the series can be reasonably found. This is the case if the general term is a quotient whose denominator can be factored into linear factors. The whole quotient is expressed as partial fractions, and then by means of this lemma it soon becomes clear whether we can find the partial sum.

Example 1. *Find the partial sum of the series* $1+\frac{1}{3}+\frac{1}{6}+\frac{1}{10}+\frac{1}{15}+\frac{1}{21}+\cdots$, *whose general term is* $2/\left(x^2+x\right)$.

This general term can be expressed as

$$\frac{2}{x} - \frac{2}{x+1}.$$

It follows that the partial sum is

$$2S.\frac{1}{x} - 2S.\frac{1}{x+1}.$$

By the previous lemma we have that this is equal to

$$2 - \frac{2}{x+1} = \frac{2x}{x+1}.$$

Hence, if $x = 4$, then $\frac{8}{5} = 1 + \frac{1}{3} + \frac{1}{6} + \frac{1}{10}$.

Example 2. *Find the partial sum of the series* $\frac{1}{5}, \frac{1}{21}, \frac{1}{45}, \frac{1}{77}, \frac{1}{117}, \ldots$, *whose general term is* $1/\left(4x^2+4x-3\right)$.

Since the denominator of the general term has the factors $(2x-1)$ and $(2x+3)$, the general term can be expressed as

$$\frac{1}{4}\cdot\frac{1}{2x-1}-\frac{1}{4}\cdot\frac{1}{2x+3}=\frac{1}{8}\cdot\frac{1}{x-\frac{1}{2}}-\frac{1}{8}\cdot\frac{1}{x+\frac{3}{2}}.$$

But then

$$S.\frac{1}{x-\frac{1}{2}}=S.\frac{1}{x+\frac{1}{2}}+2-\frac{1}{x+\frac{1}{2}}$$

and

$$S.\frac{1}{x+\frac{1}{2}}=S.\frac{1}{x+\frac{3}{2}}+\frac{2}{3}-\frac{1}{x+\frac{3}{2}},$$

so that

$$S.\frac{1}{x-\frac{1}{2}}-S.\frac{1}{x+\frac{3}{2}}=2+\frac{2}{3}-\frac{1}{x+\frac{1}{2}}-\frac{1}{x+\frac{3}{2}}.$$

When this is divided by eight, we have the desired partial sum:

$$\frac{1}{4}+\frac{1}{12}-\frac{1}{8x+4}-\frac{1}{8x+12}=\frac{x}{4x+2}+\frac{x}{3(4x+6)}=\frac{x(4x+5)}{3(2x+1)(2x+3)}.$$

70. General terms that have the form of binomial coefficients deserve special notice. We will find the partial sums of series whose numerators are 1 and whose denominators are binomial coefficients. Hence, a series whose:

general term is	has partial sum
$\frac{1\cdot 2}{x(x+1)}$	$\frac{2}{1}-\frac{2}{x+1}$,
$\frac{1\cdot 2\cdot 3}{x(x+1)(x+2)}$	$\frac{3}{2}-\frac{1\cdot 3}{(x+1)(x+2)}$,
$\frac{1\cdot 2\cdot 3\cdot 4}{x(x+1)(x+2)(x+3)}$	$\frac{4}{3}-\frac{1\cdot 2\cdot 4}{(x+1)(x+2)(x+3)}$,
$\frac{1\cdot 2\cdot 3\cdot 4\cdot 5}{x(x+1)(x+2)(x+3)(x+4)}$	$\frac{5}{4}-\frac{1\cdot 2\cdot 3\cdot 5}{(x+1)(x+2)(x+3)(x+4)}$,

and so forth. From this the law by which these expressions proceed is obvious. However, the partial sum that corresponds to the general term $1/x$ is not obtained, since it cannot be expressed in finite form.

71. Since the partial sum has x terms if the index is x, it is clear that if we let the index become infinite, we obtain the sum of these infinite series.

In this case, in the expressions just found, the later terms will vanish due to the denominators becoming infinite.

Hence, these infinite series have finite sums, as follows:

$$1+\frac{1}{3}+\frac{1}{6}+\frac{1}{10}+\frac{1}{15}+\cdots=\frac{2}{1},$$

$$1+\frac{1}{4}+\frac{1}{10}+\frac{1}{20}+\frac{1}{35}+\cdots=\frac{3}{2},$$

$$1+\frac{1}{5}+\frac{1}{15}+\frac{1}{35}+\frac{1}{70}+\cdots=\frac{4}{3},$$

$$1+\frac{1}{6}+\frac{1}{21}+\frac{1}{56}+\frac{1}{126}+\cdots=\frac{5}{4},$$

$$1+\frac{1}{7}+\frac{1}{28}+\frac{1}{84}+\frac{1}{210}+\cdots=\frac{6}{5},$$

and so forth. Therefore, every series whose partial sum we know can be continued to infinity, and the sum can be exhibited by letting $x=\infty$, provided that the sum is finite; this will be the case if in the partial sum the power of x is the same in both the numerator and denominator.

3

On the Infinite and the Infinitely Small

72. Since every quantity, no matter how large, can always be increased, and there is no obstacle to adding to a given quantity another like quantity, it follows that every quantity can be increased without limit. Furthermore, there is no quantity so large that a larger one cannot be conceived, and so there is no doubt that *every quantity can be increased to infinity.* If there is someone who would deny this, he would have to give some quantity that cannot be increased, and so he needs to give a quantity to which nothing can be added. This is absurd, and even the idea of quantity rules out this possibility. He must necessarily concede that every quantity can always be increased without limit, that is, it can be increased to infinity.

73. For each kind of quantity this becomes even clearer. No one can easily defend himself if he declares that the series of natural numbers, $1, 2, 3, 4, 5, 6, \ldots$ has a limit beyond which it cannot be continued. Indeed, there is no such number to which 1 cannot be added to obtain the following number, which is greater. Hence, the series of natural numbers continues without limit, nor is it possible to come to some greatest number beyond which there is no greater number. In like manner the straight line cannot be extended to such a point that it cannot be extended further. By this it is clear that both the integers and the line can be increased to infinity. No matter what kind of quantity it may be, we should understand that every quantity, no matter how large, can always be made greater and greater, and thus increased without limit, that is, increased to infinity.

74. Although these things are clear enough, so that anyone who would deny them must contradict himself, still, this theory of the infinite has been so obfuscated by so many difficulties and even involved in contradictions by many who have tried to explain it, that no way is open by which they may extricate themselves. From the fact that a quantity can be increased to infinity, some have concluded that there is actually an infinite quantity, and they have described it in such a way that it cannot be increased. In this way they overturn the very idea of quantity, since they propose a quantity of such a kind that it cannot be increased. Furthermore, those who admit such an infinity contradict themselves; when they put an end to the capacity a quantity has of being increased, they simultaneously deny that the quantity can be increased without limit, since these two statements come to the same thing. Thus while they admit an infinite quantity, they also deny it. Indeed, if a quantity cannot be increased without limit, that is to infinity, then certainly no infinite quantity can exist.

75. Hence, from the fact that every quantity can be increased to infinity, it seems to follow that there is no infinite quantity. A quantity increased continuously by increments does not become infinite unless it shall have already increased without limit. However, that which must increase without limit cannot be conceived of as having already become infinite. Nevertheless, not only is it possible to give a quantity of this kind, to which increments are added without limit, a certain character, and with due care to introduce it into calculus, as we shall soon see at length, but also there exist real cases, at least they can be conceived, in which an infinite number actually exists. Thus, if there are things that are infinitely divisible, as many philosophers have held to be the case, the number of parts of which this thing is constituted is really infinite. Indeed, if it be claimed that the number is finite, then the thing is not really infinitely divisible. In a like manner, if the whole world were infinite, as many have held, then the number of bodies making up the world would certainly not be finite, and would hence be infinite.

76. Although there seems to be a contradiction here, if we consider it carefully we can free ourselves from all difficulties. Whoever claims that some material is infinitely divisible denies that in the continuous division of the material one ever arrives at parts so small that they can no longer be divided. Hence, this material does not have ultimate indivisible parts, since the individual particles at which one arrives by continued division must be able to be further subdivided. Therefore, whoever says, in this case, that the number of parts is infinite, also understands that the ultimate parts are indivisible; he tries to count those parts that are never reached, and hence do not exist. If some material can always be further subdivided, it lacks indivisible or absolutely simple parts. For this reason, whoever claims that some material can be infinitely subdivided denies that the material is made up of simple parts.

77. As long as we are speaking about the parts of some body or of some material, we understand not ultimate or simple parts, of which indeed there are none, but those that division really produces. Then, since by hypothesis we admit material that is infinitely divisible, even a very small particle of material can be dissected into many parts, but no number can be given that is so large that a greater number of parts cut from that particle cannot be exhibited. Hence the number of parts, indeed not ultimate parts, but those that are still further divisible, that make up each body, is greater than any number that can be given. Likewise, if the whole world is infinite, the number of bodies making up the world is greater than any assignable number. Since this is not a finite number, it follows that an infinite number and a number greater than any assignable number are two ways of saying the same thing.

78. Anyone who has gathered from this discussion any insight into the infinite divisibility of matter will suffer none of the difficulties that people commonly assign to this opinion. Nor will he be forced to admit anything contrary to sound reasoning. On the other hand, anyone who denies that matter is infinitely divisible will find himself in serious difficulties from which he will in no way be able to extricate himself. These ultimate particles are called by some *atoms*, by others *monads* or *simple beings*. The reason why these ultimate particles admit no further division could be for two possible reasons. The first is that they have no extension; the second is that although they have extension, they are so hard and impenetrable that no force is sufficient to dissect them. Whichever choice is made will lead to equally difficult positions.

79. Suppose that ultimate particles lack any extension, so that they lack any further parts: By this explanation the idea of simple beings is nicely saved. However, it is impossible to conceive how a body can be constituted by a finite number of particles of this sort. Suppose that a cubic foot of matter is made up of a thousand simple beings of this kind, and that it is actually cut up into one thousand pieces. If these pieces are equal, they will each be one cubic finger; if they are not equal, some will be larger, some smaller. One cubic finger will be a simple being, and we will be faced with a great contradiction, unless by chance we want to say that there is one simple being and the rest of the space is empty. In this way the continuity of the body is denied, except that those philosophers completely banished any vacuum from the world. If someone should object that the number of simple beings contained in a cubic foot of matter is much more than a thousand, absolutely nothing is gained. Any difficulty that follows from the number one thousand will remain with any other number, no matter how large. The inventor of the monad, a very acute man, LEIBNIZ, probed this problem deeply, and finally decided that matter is infinitely divisible. Hence, it is not possible to arrive at a monad before the body is actually infinitely divided. By this very fact, the existence of simple beings that make up a body is completely refuted. He who denies that bodies are made up of simple beings and he who claims that bodies are infinitely divisible are both saying the same thing.

80. Nor is their position any better if they say that the ultimate particles of a body are indeed extended, but because of the hardness they cannot be broken apart. Since they admit extension in the ultimate particles, they hold that the particles are composite. Whether or not they can be separated from each other makes little difference, since they can assign no cause that explains this hardness. For the most part, however, those who deny the infinite divisibility of matter seem to have sufficiently felt the difficulties of this latter position, since usually they cling to the former idea. But they cannot escape these difficulties except with a few trivial metaphysical distinctions, which generally strive to keep us from trusting the consequences that follow from mathematical principles. Nor should they admit that simple parts have dimensions. In the first place they should have demonstrated that these ultimate parts, of which a determined number make up a body, have no extension.

81. Since they can find no way out of this labyrinth, nor can they meet the objections in a suitable way, they flee to distinctions, and to the objections they reply with arguments supplied by the senses and the imagination. In this situation one should rely solely on the intellect, since the senses and arguments depending on them frequently are fallacious. Pure intellect admits the possibility that one thousandth part of a cubic foot of matter might lack all extension, while this seems absurd to the imagination. That which frequently deceives the senses may be true, but it can be decided by no one except mathematicians. Indeed, mathematics defends us in particular against errors of the senses and teaches about objects that are perceived by the senses, sometimes correctly, and sometimes only in appearance. This is the safest science, whose teaching will save those who follow it from the illusions of the senses. It is far removed from those responses by which metaphysicians protect their doctrine and thus rather make it more suspect.

82. But let us return to our proposition. Even if someone denies that infinite numbers really exist in this world, still in mathematical speculations there arise questions to which answers cannot be given unless we admit an infinite number. Thus, if we want the sum of all the numbers that make up the series $1 + 2 + 3 + 4 + 5 + \cdots$, since these numbers progress with no end, and the sum increases, it certainly cannot be finite. By this fact it becomes infinite. Hence, this quantity is so large that it is greater than any finite quantity and cannot not be infinite. To designate a quantity of this kind we use the symbol ∞, by which we mean a quantity greater than any finite or assignable quantity. Thus, when a parabola needs to be defined in such a way that it is said to be an infinitely long ellipse, we can correctly say that the axis of the parabola is an infinitely long straight line.

83. This theory of the infinite will be further illustrated if we discuss that which mathematicians call the infinitely small. There is no doubt that any quantity can be diminished until it all but vanishes and then goes to nothing. But an infinitely small quantity is nothing but a vanishing quantity, and so it is really equal to 0. There is also a definition of the infinitely small quantity as that which is less than any assignable quantity. If a quantity is so small that it is less than any assignable quantity, then it cannot not be 0, since unless it is equal to 0 a quantity can be assigned equal to it, and this contradicts our hypothesis. To anyone who asks what an infinitely small quantity in mathematics is, we can respond that it really is equal to 0. There is really not such a great mystery lurking in this idea as some commonly think and thus have rendered the calculus of the infinitely small suspect to so many. In the meantime any doubts that may remain will be removed in what follows, where we are going to treat this calculus.

84. Since we are going to show that an infinitely small quantity is really zero, we must first meet the objection of why we do not always use the same symbol 0 for infinitely small quantities, rather than some special ones. Since all nothings are equal, it seems superfluous to have different signs to designate such a quantity. Although two zeros are equal to each other, so that there is no difference between them, nevertheless, since we have two ways to compare them, either arithmetic or geometric, let us look at quotients of quantities to be compared in order to see the difference. The arithmetic ratio between any two zeros is an equality. This is not the case with a geometric ratio. We can easily see this from this geometric proportion $2 : 1 = 0 : 0$, in which the fourth term is equal to 0, as is the third. From the nature of the proportion, since the first term is twice the second, it is necessary that the third is twice the fourth.

85. These things are very clear, even in ordinary arithmetic. Everyone knows that when zero is multiplied by any number, the product is zero and that $n \cdot 0 = 0$, so that $n : 1 = 0 : 0$. Hence, it is clear that any two zeros can be in a geometric ratio, although from the perspective of arithmetic, the ratio is always of equals. Since between zeros any ratio is possible, in order to indicate this diversity we use different notations on purpose, especially when a geometric ratio between two zeros is being investigated. In the calculus of the infinitely small, we deal precisely with geometric ratios of infinitely small quantities. For this reason, in these calculations, unless we use different symbols to represent these quantities, we will fall into the greatest confusion with no way to extricate ourselves.

86. If we accept the notation used in the analysis of the infinite, then dx indicates a quantity that is infinitely small, so that both $dx = 0$ and

$a\,dx = 0$, where a is any finite quantity. Despite this, the geometric ratio $a\,dx : dx$ is finite, namely $a : 1$. For this reason these two infinitely small quantities dx and $a\,dx$, both being equal to 0, cannot be confused when we consider their ratio. In a similar way, we will deal with infinitely small quantities dx and dy. Although these are both equal to 0, still their ratio is not that of equals. Indeed, the whole force of differential calculus is concerned with the investigation of the ratios of any two infinitely small quantities of this kind. The application of these ratios at first sight might seem to be minimal. Nevertheless, it turns out to be very great, which becomes clearer with each passing day.

87. Since the infinitely small is actually nothing, it is clear that a finite quantity can neither be increased nor decreased by adding or subtracting an infinitely small quantity. Let a be a finite quantity and let dx be infinitely small. Then $a + dx$ and $a - dx$, or, more generally, $a \pm n\,dx$, are equal to a. Whether we consider the relation between $a \pm n\,dx$ and a as arithmetic or as geometric, in both cases the ratio turns out to be that between equals. The arithmetic ratio of equals is clear: Since $n\,dx = 0$, we have

$$a \pm n\,dx - a = 0.$$

On the other hand, the geometric ratio is clearly of equals, since

$$\frac{a \pm n\,dx}{a} = 1.$$

From this we obtain the well-known rule that *the infinitely small vanishes in comparison with the finite and hence can be neglected.* For this reason the objection brought up against the analysis of the infinite, that it lacks geometric rigor, falls to the ground under its own weight, since nothing is neglected except that which is actually nothing. Hence with perfect justice we can affirm that in this sublime science we keep the same perfect geometric rigor that is found in the books of the ancients.

88. Since the infinitely small quantity dx is actually equal to 0, its square dx^2, cube dx^3, and any other dx^n, where n is a positive exponent, will be equal to 0, and hence in comparison to a finite quantity will vanish. However, even the infinitely small quantity dx^2 will vanish when compared to dx. The ratio of $dx \pm dx^2$ to dx is that of equals, whether the comparison is arithmetic or geometric. There is no doubt about the arithmetic; in the geometric comparison,

$$dx \pm dx^2 : dx = \frac{dx \pm dx^2}{dx} = 1 \pm dx = 1.$$

In like manner we have $dx \pm dx^3 = dx$ and generally $dx \pm dx^{n+1} = dx$, provided that n is positive. Indeed, the geometric ratio $dx \pm dx^{n+1} : dx$

equals $1 + dx^n$, and since $dx^n = 0$, the ratio is that of equals. Hence, if we follow the usage of exponents, we call dx infinitely small of the first order, dx^2 of the second order, dx^3 of the third order, and so forth. It is clear that in comparison with an infinitely small quantity of the first order, those of higher order will vanish.

89. In a similar way it is shown that an infinitely small quantity of the third and higher orders will vanish when compared with one of the second order. In general, an infinitely small quantity of any higher order vanishes when compared with one of lower order. Hence, if m is less than n, then

$$a\,dx^m + b\,dx^n = a\,dx^m,$$

since dx^n vanishes compared with dx^m, as we have shown. This is true also with fractional exponents; dx vanishes compared with $\sqrt{dx}$ or $dx^{\frac{1}{2}}$, so that

$$a\sqrt{dx} + b\,dx = a\sqrt{dx}.$$

Even if the exponent of dx is equal to 0, we have $dx^0 = 1$, although $dx = 0$. Hence the power dx^n is equal to 1 if $n = 0$, and from being a finite quantity becomes infinitely small if n is greater than 0.

Therefore, there exist an infinite number of orders of infinitely small quantities. Although all of them are equal to 0, still they must be carefully distinguished one from the other if we are to pay attention to their mutual relationships, which has been explained through a geometric ratio.

90. Once we have established the concept of the infinitely small, it is easier to discuss the properties of infinity, or the infinitely large. It should be noted that the fraction $1/z$ becomes greater the smaller the denominator z becomes. Hence, if z becomes a quantity less than any assignable quantity, that is, infinitely small, then it is necessary that the value of the fraction$1/z$ becomes greater than any assignable quantity and hence infinite. For this reason, if 1 or any other finite quantity is divided by something infinitely small or 0, the quotient will be infinitely large, and thus an infinite quantity. Since the symbol ∞ stands for an infinitely large quantity, we have the equation

$$\frac{a}{dx} = \infty.$$

The truth of this is clear also when we invert:

$$\frac{a}{\infty} = dx = 0.$$

Indeed, the larger the denominator z of the fraction a/z becomes, the smaller the value of the fraction becomes, and if z becomes an infinitely large quantity, that is $z = \infty$, then necessarily the value of the fraction a/∞ becomes infinitely small.

91. Anyone who denies either of these arguments will find himself in great difficulties, with the necessity of denying even the most certain principles of analysis. If someone claims that the fraction $a/0$ is finite, for example equal to b, then when both parts of the equation are multiplied by the denominator, we obtain $a = 0 \cdot b$. Then the finite quantity b multiplied by zero produces a finite a, which is absurd. Much less can the value b of the fraction $a/0$ be equal to 0; in no way can 0 multiplied by 0 produce the quantity a. Into the same absurdity will fall anyone who denies that $a/\infty = 0$, since then he would be saying that $a/\infty = b$, a finite quantity. From the equation $a/\infty = b$ it would legitimately follow that $\infty = a/b$, but from this we conclude that the value of the fraction a/b, whose numerator and denominator are both finite quantities, is infinitely large, which of course is absurd. Nor is it possible that the values of the fractions $a/0$ and a/∞ could be complex, since the value of a fraction whose numerator is finite and whose denominator is complex cannot be either infinitely large or infinitely small.

92. An infinitely large quantity, to which we have been led through this consideration, and which is treated only in the analysis of the infinite, can best be defined by saying that an infinitely large quantity is the quotient that arises from the division of a finite quantity by an infinitely small quantity. Conversely, we can say that an infinitely small quantity is a quotient that arises from division of a finite quantity by an infinitely large quantity. Since we have a geometric proportion in which an infinitely small quantity is to a finite quantity as a finite quantity is to an infinitely large quantity, it follows that an infinite quantity is infinitely greater than a finite quantity, just as a finite quantity is infinitely greater than an infinitely small quantity. Hence, statements of this sort, which disturb many, should not be rejected, since they rest on most certain principles. Furthermore, from the equation $a/0 = \infty$, it can follow that zero multiplied by an infinitely large quantity produces a finite quantity, which would seem strange were it not the result of a very clear deduction.

93. Just as when we compare infinitely small quantities by a geometric ratio, we can find very great differences, so when we compare infinitely large quantities the difference can be even greater, since they differ not only by geometric ratios, but also by arithmetic. Let A be an infinite quantity that is obtained from division of a finite quantity a by the infinitely small dx, so that $a/dx = A$. Likewise $2a/dx = 2A$ and $na/dx = nA$. Now, since nA is an infinite quantity, it follows that the ratio between two infinitely large quantities can have any value. Hence, if an infinite quantity is either multiplied or divided by a finite number, the result will be an infinite quantity. Nor can it be denied that infinite quantities can be further augmented. It is easily seen that if the geometric ratio that holds between two infinite quantities shows them to be unequal, then even less will an arithmetic ratio show them to be equal, since their difference will always be infinitely large.

94. Although there are some for whom the idea of the infinite, which we use in mathematics, seems to be suspect, and for this reason think that analysis of the infinite is to be rejected, still even in the trivial parts of mathematics we cannot do without it. In arithmetic, where the theory of logarithms is developed, the logarithm of zero is said to be both negative and infinite. There is no one in his right mind who would dare to say that this logarithm is either finite or even equal to zero. In geometry and trigonometry this is even clearer. Who is there who would ever deny that the tangent or the secant of a right angle is infinitely large? Since the rectangle formed by the tangent and the cotangent has an area equal to the square of the radius, and the cotangent of a right angle is equal to 0, even in geometry it has to be admitted that the product of zero and infinity can be finite.

95. Since a/dx is an infinite quantity A, it is clear that the quantity A/dx will be a quantity infinitely greater than the quantity A. This can be seen from the proportion $a/dx : A/dx = a : A$, that is, as a finite number to one infinitely large. There are relations of this kind between infinitely large quantities, so that some can be infinitely greater than others. Thus, a/dx^2 is a quantity infinitely greater than a/dx; if we let $a/dx = A$, then $a/dx^2 = A/dx$. In a similar way a/dx^3 is an infinite quantity infinitely greater than a/dx^2, and so is infinitely greater than a/dx. We have, therefore an infinity of grades of infinity, of which each is infinitely greater than its predecessor. If the number m is just a little bit greater than n, then a/dx^m is an infinite quantity infinitely greater than the infinite quantity a/dx^n.

96. Just as with infinitely small quantities there are geometric ratios indicating inequalities, but arithmetic ratios always indicate equality, so with infinitely large quantities we have geometric ratios indicating equality, but whose arithmetic ratios still indicate inequality. If a and b are two finite quantities, then the geometric ratio of two infinite quantities $a/dx + b$ and a/dx indicates that the two are equal; the quotient of the first by the second is equal to $1 + b\,dx/a = 1$, since $dx = 0$. However, if they are compared arithmetically, due to the difference b, the ratio indicates inequality. In a similar way, the geometric ratio of $a/dx^2 + a/dx$ to a/dx^2 indicates equality; expressing the ratio, we have $1 + dx = 1$, since $dx = 0$. On the other hand, the difference is a/dx, and so this is infinite. It follows that when we consider geometric ratios, an infinitely large quantity of a lower grade will vanish when compared to an infinitely large quantity of a higher grade.

97. Now that we have been warned about the grades of infinities, we will soon see that it is possible not only for the product of an infinitely large quantity and an infinitely small quantity to produce a finite quantity, as we have already seen, but also that a product of this kind can also be either infinitely large or infinitely small. Thus, if the infinite quantity a/dx is multiplied by the infinitely small dx, the product will be equal to the finite a. However, if a/dx is multiplied by the infinitely small dx^2 or dx^3 or another of higher order, the product will be $a\,dx$, $a\,dx^2$, $a\,dx^3$, and so forth, and so it will be infinitely small. In the same way, we understand that if the infinite quantity a/dx^2 is multiplied by the infinitely small dx, then the product will be infinitely large. In general, if a/dx^n is multiplied by $b\,dx^m$, the product $ab\,dx^{m-n}$ will be infinitely small if m is greater than n; it will be finite if m equals n; it will be infinitely large if m is less than n.

98. Both infinitely small and infinitely large quantities often occur in series of numbers. Since there are finite numbers mixed in these series, it is clearer than daylight, how, according to the laws of continuity, one passes from finite quantities to infinitely small and to infinitely large quantities. First let us consider the series of natural numbers, continued both forward and backward:

$$\ldots,\ -4,\ -3,\ -2,\ -1,\ +0,\ +1,\ +2,\ +3,\ +4,\ \ldots.$$

By continuously decreasing, the numbers approach 0, that is, the infinitely small. Then they continue further and become negative. From this we understand that the positive numbers decrease, passing through 0 to increasing negative numbers. However, if we consider the squares of the numbers, since they are all positive,

$$\ldots,\ +16,\ +9,\ +4,\ +1,\ +0,\ +1,\ +4,\ +9,\ +16,\ \ldots,$$

we have 0 as the transition number from the decreasing positive numbers to the increasing positive numbers. If all of the signs are changed, then 0 is again the transition from decreasing negative numbers to increasing negative numbers.

99. If we consider the series with general term $\sqrt{x}$, which is continued both forwards and backwards, we have

$$\ldots,\ +\sqrt{-3},\ +\sqrt{-2},\ +\sqrt{-1},\ +0,\ +\sqrt{1},\ +\sqrt{2},\ +\sqrt{3},\ +\sqrt{4},\ \ldots,$$

and from this it is clear that 0 is a kind of limit through which real quantities pass to the complex.

If these terms are considered as points on a curve, it is seen that if they are positive and decrease so that they eventually vanish, then continuing further, they become either negative, or positive again, or even complex. The same happens if the points were first negative, then also vanish, and if they continue further, become either positive, negative, or complex. Many examples of phenomena of this kind are found in the theory of plane curves, treated in a preceding book.[1]

100. In the same way infinite terms often occur in series. Thus, in the harmonic series, whose general term is $1/x$, the term corresponding to the index $x = 0$ is the infinite term $1/0$. The whole series is as follows:

$$\ldots, \quad -\frac{1}{4}, \quad -\frac{1}{3}, \quad -\frac{1}{2}, \quad -\frac{1}{1}, \quad +\frac{1}{0}, \quad +\frac{1}{1}, \quad +\frac{1}{2}, \quad +\frac{1}{3}, \quad \ldots.$$

Going from right to left the terms increase, so that $1/0$ is infinitely large. Once it has passed through, the terms become decreasing and negative. Hence, an infinitely large quantity can be thought of as some kind of limit, passing through which positive numbers become negative and vice versa. For this reason it has seemed to many that the negative numbers can be thought of as greater than infinity, since in this series the terms continuously increase, and once they have reached infinity, they become negative. However, if we consider the series whose general term is $1/x^2$, then after passing through infinity, the terms become positive again,

$$\ldots, \quad +\frac{1}{9}, \quad +\frac{1}{4}, \quad +\frac{1}{1}, \quad +\frac{1}{0}, \quad +\frac{1}{1}, \quad +\frac{1}{4}, \quad +\frac{1}{9}, \quad \ldots,$$

and no one would say that these are greater than infinity.

101. Frequently, in a series an infinite term will constitute a limit separating real terms from complex, as occurs in the following series, whose general term is $1/\sqrt{x}$:

$$\ldots, \quad +\frac{1}{\sqrt{-3}}, \quad +\frac{1}{\sqrt{-2}}, \quad +\frac{1}{\sqrt{-1}}, \quad +\frac{1}{0}, \quad +\frac{1}{\sqrt{1}}, \quad +\frac{1}{\sqrt{2}}, \quad +\frac{1}{\sqrt{3}}, \quad \ldots.$$

From this it does not follow that complex numbers are greater than infinity, since from the series previously discussed,

$$\ldots, \quad +\sqrt{-3}, \quad +\sqrt{-2}, \quad +\sqrt{-1}, \quad +0, \quad +\sqrt{1}, \quad +\sqrt{2}, \quad +\sqrt{3}, \quad \ldots,$$

it would equally follow that the complex numbers are less than zero. It is possible to show a change from real terms to complex, where the limit is neither 0 nor ∞, for example if the general term is $1 + \sqrt{x}$. In these cases, due to the irrationality, each term has two values. In the limit between real and complex numbers the two values always come together as equals. Nevertheless, whenever there are terms that are first positive and then become negative, the transition is always through a limit that is infinitely small or infinitely large. This is all due to the law of continuity, which is most clearly seen through plane curves.

---

[1] *Introduction*, Book II.

102. From the summation of infinite series we can gather many results that both further illustrate this theory of the infinite and also aid in answering doubts that frequently arise in this material. In the first place, if the series has equal terms, such as

$$1 + 1 + 1 + 1 + 1 + \cdots ,$$

which is continued to infinity, there is no doubt that the sum of all of these terms is greater than any assignable number. For this reason it must be infinite. We confirm this by considering its origin in the expansion of the fraction

$$\frac{1}{1-x} = 1 + x + x^2 + x^3 + \cdots .$$

If we let $x = 1$, then

$$\frac{1}{1-1} = 1 + 1 + 1 + 1 + \cdots ,$$

so that the sum is equal to

$$\frac{1}{1-1} = \frac{1}{0} = \infty.$$

103. Although there can be no doubt that when the same finite number is added an infinite number of times the sum should be infinite, still, the general infinite series that originates from the fraction

$$\frac{1}{1-x} = 1 + x + x^2 + x^3 + x^4 + x^5 + \cdots$$

seems to labor under most serious difficulties. If for x we successively substitute the numbers $1, 2, 3, 4, \ldots$, we obtain the following series with their sums:

A. $1 + 1 + 1 + 1 + 1 + \cdots = \frac{1}{1-1} = \infty,$

B. $1 + 2 + 4 + 8 + 16 + \cdots = \frac{1}{1-2} = -1,$

C. $1 + 3 + 9 + 27 + 81 + \cdots = \frac{1}{1-3} = -\frac{1}{2},$

D. $1 + 4 + 16 + 64 + 256 + \cdots = \frac{1}{1-4} = -\frac{1}{3},$

and so forth. Since each term of series B, except for the first, is greater than the corresponding term of series A, the sum of series B must be much more than the sum of series A. Nevertheless, this calculation shows that series A has an infinite sum, while series B has a negative sum, which is less than zero, and this is beyond comprehension. Even less can we reconcile with ordinary ideas the results of this and the following series C, D, and so forth, which have negative sums while all of the terms are positive.

104. For this reason, the opinion suggested above, namely, that negative numbers might sometimes be considered greater than the infinite, that is, more than infinity, might seem to be more probable. Since it is also true that when decreasing numbers go beyond zero they become negative, a distinction has to be made between negative numbers like $-1, -2, -3, \ldots$ and negative numbers like

$$\frac{+1}{-1} \quad \frac{+2}{-1} \quad \frac{+3}{-1}, \quad \ldots,$$

the former being less than zero and the latter being greater than the infinite. Even with this agreement, the difficulty is not eliminated, as is suggested by the following series:

$$1 + 2x + 3x^2 + 4x^3 + 5x^4 + \cdots = \frac{1}{(1-x)^2},$$

from which we obtain the following series:

A. $1 + 2 + 3 + 4 + 5 + \cdots = \frac{1}{(1-1)^2} = \frac{1}{0} = \infty,$

B. $1 + 4 + 12 + 32 + 80 + \cdots = \frac{1}{(1-2)^2} = 1.$

Now, every term of series B is greater than the corresponding term of series A, except for the first term, and insofar as the sum of series A is infinite, while the sum of series B is equal to 1, which is only the first term, the suggested principle is no explanation at all.

105. Since if we were to deny that

$$-1 = \frac{+1}{-1} \qquad \text{and} \qquad \frac{+a}{-b} = \frac{-a}{+b},$$

the very firmest foundations of analysis would collapse, the previously suggested explanation is not to be admitted. We ought rather to deny that the sums that the general formula supplied are the true sums. Since these series are derived by continual division, and while the remainders are divided further, the remainders always grow larger the longer we continue, so that the remainder can never be neglected. Even less can the last remainder, that is, that divided by an infinitesimal, be omitted, since it is infinite. Since we did not observe this in the previous series where the remainder became zero, it should not be surprising that those sums led to absurd results. Since this response is derived from the very origin of the series itself, it is most true and it removes all doubt.

106. In order that this may be clarified, let us examine the development of the fraction $1/(1-x)$ in the first finite number of terms. Hence we have

$$\frac{1}{1-x} = 1 + \frac{x}{1-x},$$

$$\frac{1}{1-x} = 1 + x + \frac{x^2}{1-x},$$

$$\frac{1}{1-x} = 1 + x + x^2 + \frac{x^3}{1-x},$$

$$\frac{1}{1-x} = 1 + x + x^2 + x^3 + \frac{x^4}{1-x},$$

and so forth. If someone wishes to say that the finite series $1+x+x^2+x^3$ has a sum equal to $1/(1-x)$, then he is in error by the quantity $x^4/(1-x)$; if he should say that the sum of the series $1+x+x^2+x^3+\cdots+x^{1000}$ is $1/(1-x)$, then his error is equal to $x^{1001}/(1-x)$. If x happens to be greater than 1, this error is very large.

107. From this we see that he who would say that when this same series is continued to infinity, that is,

$$1 + x + x^2 + x^3 + \cdots + x^\infty,$$

and that the sum is $1/(1-x)$, then his error would be $x^{\infty+1}/(1-x)$, and if $x > 1$, then the error is indeed infinite. At the same time, however, this same argument shows why the series $1+x+x^2+x^3+x^4+\cdots$, continued to infinity, has a true sum of $1/(1-x)$, provided that x is a fraction less than 1. In this case the error $x^{\infty+1}$ is infinitely small and hence equal to zero, so that it can safely be neglected. Thus if we let $x = \frac{1}{2}$, then in truth

$$1 + \frac{1}{2} + \frac{1}{4} + \frac{1}{8} + \frac{1}{16} + \cdots = \frac{1}{1-\frac{1}{2}} = 2.$$

In a similar way, the rest of the series in which x is a fraction less than 1 will have a true sum in the way we have indicated.

108. This same answer is valid for the sum of divergent series in which the signs alternate between $+$ and $-$, which ordinarily is given by the same formula, but with the sign of x changed to negative. Since we have

$$\frac{1}{1+x} = 1 - x + x^2 - x^3 + x^4 - x^5 + \cdots,$$

if we did not express the final remainder, we would have

A. $1 - 1 + 1 - 1 + 1 - 1 + \cdots = \frac{1}{2}$,

B. $1-2+4-8+16-32+\cdots=\frac{1}{3}$,

C. $1-3+9-27+81-243+\cdots=\frac{1}{4}$.

It is clear that the sum of series B cannot be equal to $\frac{1}{3}$, since the more terms we actually sum, the farther away the result gets from $\frac{1}{3}$. But the sum of any series ought to be a limit the closer to which the partial sums should approach, the more terms are added.

109. From this we conclude that series of this kind, which are called divergent, have no fixed sums, since the partial sums do not approach any limit that would be the sum for the infinite series. This is certainly a true conclusion, since we have shown the error in neglecting the final remainder. However, it is possible, with considerable justice, to object that these sums, even though they seem not to be true, never lead to error. Indeed, if we allow them, then we can discover many excellent results that we would not have if we rejected them out of hand. Furthermore, if these sums were really false, they would not consistently lead to true results; rather, since they differ from the true sum not just by a small difference, but by infinity, they should mislead us by an infinite amount. Since this does not happen, we are left with a most difficult knot to unravel.

110. I say that the whole difficulty lies in the name *sum*. If, as is commonly the case, we take the *sum* of a series to be the aggregate of all of its terms, actually taken together, then there is no doubt that only infinite series that converge continually closer to some fixed value, the more terms we actually add, can have sums. However, divergent series, whose terms do not decrease, whether their signs + and − alternate or not, do not really have fixed sums, supposing we use the word *sum* for the aggregate of all of the terms. Consider these cases that we have recalled, with erroneous sums, for example the finite expression $1/(1-x)$ for the infinite series $1+x+x^2+x^3+\cdots$. The truth of the matter is this, not that the expression is the sum of the series, but that the series is derived from the expression. In this situation the name *sum* could be completely omitted.

111. These inconveniences and apparent contradictions can be avoided if we give the word *sum* a meaning different from the usual. Let us say that the *sum* of any infinite series is a finite expression from which the series can be derived. In this sense, the true sum of the infinite series $1+x+x^2+x^3+\cdots$ is $1/(1-x)$, since this series is derived from the fraction, no matter what value is substituted for x. With this understanding, if the series is convergent, the new definition of sum agrees with the usual definition. Since divergent series do not have a sum, properly speaking, there is no real difficulty which arises from this new meaning. Finally, with the aid of this definition we can keep the usefulness of divergent series and preserve their reputations.

4
On the Nature of Differentials of Each Order

112. In the first chapter we saw that if the variable quantity x received an increment equal to ω, then from this each function of x obtained an increment that can be expressed as $P\omega + Q\omega^2 + R\omega^3 + \cdots$, and this expression may be finite or it may go to infinity. Hence the function y, when we write $x + \omega$ for x, takes the following form:

$$y^{\mathrm{I}} = y + P\omega + Q\omega^2 + R\omega^3 + S\omega + \cdots .$$

When the previous value of y is subtracted, there remains the difference of the function y, which we express as

$$\Delta y = P\omega + Q\omega^2 + R\omega^3 + S\omega^4 + \cdots .$$

Since the subsequent value of x is $x^{\mathrm{I}} = x + \omega$, we have the difference of x, namely, $\Delta x = \omega$. The letters $P, Q, R, \ldots$ represent functions of x, depending on y, which we found in the first chapter.

113. Therefore, with whatever increment ω the variable quantity x is increased, at the same time it is possible to define the increase that accrues to y, the function of x, provided that we can define the functions $P, Q, R, S, \ldots$ for any function y. In this chapter, and in all of the analysis of the infinite, the increment ω by which we let the variable x increase will be infinitely small, so that it vanishes; that is, it is equal to 0. Hence it is clear that the increase, or the difference, of the function y will also be infinitely small. With this hypothesis, each term of the expression

$$P\omega + Q\omega^2 + R\omega^3 + S\omega^4 + \cdots$$

will vanish when compared with its predecessor (paragraph 88 and following), so that only $P\omega$ will remain. For this reason, in the present case where ω is infinitely small, the difference of y, Δy, is equal to $P\omega$.

114. The analysis of the infinite, which we begin to treat now, is nothing but a special case of the method of differences, explained in the first chapter, wherein the differences are infinitely small, while previously the differences were assumed to be finite. Hence, this case, in which the whole of analysis of the infinite is contained, should be distinguished from the method of differences. We use special names and notation for the infinitely small differences. With Leibniz we call infinitely small differences by the name *differentials.* From the discussion in the first chapter on the different orders of differences, we can easily understand the meaning of first, second, third, and so forth, differentials of any function. Instead of the symbol Δ, by which we previously indicated a difference, now we will use the symbol d, so that dy signifies the first differential of y, d^2y the second differential, d^3y the third differential, and so forth.

115. Since the infinitely small differences that we are now discussing we call *differentials*, the whole calculus by means of which differentials are investigated and applied has usually been called *differential calculus.* The English mathematicians (among whom Newton first began to develop this new branch of analysis, as did Leibniz among the Germans) use different names and symbols. They call infinitely small differences, which we call differentials, *fluxions* and sometimes *increments.* These words seem to fit better in Latin, and they signify reasonably well the things themselves. A variable quantity by continuously increasing takes on various different values, and for this reason can be thought of as being in flux, from which comes the word fluxion. This was first used by Newton for the rate of change, to designate an infinitely small increment that a quantity receives, as if, by analogy, it were flowing.

116. It might be uncivil to argue with the English about the use of words and a definition, and we might easily be defeated in a judgment about the purity of Latin and the adequacy of expression, but there is no doubt that we have won the prize from the English when it is a question of notation. For differentials, which they call fluxions, they use dots above the letters. Thus, $\dot{y}$ signifies the first fluxion of y, $\ddot{y}$ is the second fluxion, the third fluxion has three dots, and so forth. This notation, since it is arbitrary, cannot be criticized if the number of dots is small, so that the number can be recognized at a glance. On the other hand, if many dots are required, much confusion and even more inconvenience may be the result. For example, the tenth differential, or fluxion, is very inconveniently represented with ten dots, while our notation, $d^{10}y$, is very easily understood. There are cases where differentials of even much higher order, or even those of indefinite order, must be represented, and for this the English mode is completely inapt.

117. We have used both the words and notations that have been accepted in our countries; they are both more familiar and more convenient. Still, it is not beside the point that we have spoken about the English usage and notation, since those who peruse their books will need to know this if they are to be intelligible. The English are not so wedded to their ways that they refuse to read writings that use our methods. Indeed, we have read some of their works with great avidity, and have taken from them much profit. I have also often remarked that they have profited from reading works from our regions. For these reasons, although it is greatly to be desired that everywhere the same mode of expression be used, still it is not so difficult to accustom ourselves to both methods, so that we can profit from books written in their way.

118. Since up to this time we have used the letter ω to denote the difference or the increment by which the variable x is understood to increase, now we understand ω to be infinitely small, so that ω is the differential of x, and for this reason we use our method of writing $\omega = dx$. From now on, dx will be the infinitely small difference by which x is understood to increase. In like manner the differential of y we express as dy. If y is any function of x, the differential dy will indicate the increment that y receives when x changes to $x + dx$. Hence, if we substitute $x + dx$ for x in the function y and we let y^{I} be the result, then $dy = y^{\mathrm{I}} - y$, and this is understood to be the first differential, that is, the differential of the first order. Later we will consider the other differentials.

119. We must emphasize the fact that the letter d that we are using here does not denote a quantity, but is used to express the word *differential*, in the same way that the letter l is used for the word logarithm when the theory of logarithms is being discussed. In algebra we are used to using the symbol $\surd$ for a root. Hence dy does not signify, as it usually does in analysis, the product of two quantities d and y, but rather we say the differential of y. In a similar way, if we write d^2y, this is not the square of a quantity d, but it is simply a short and apt way of writing the *second differential.* Since we use the letter d in differential calculus not for some quantity, but only as a symbol, in order to avoid confusion in calculations when many different constant quantities occur, we avoid using the letter d. Just so we usually avoid the letter l to designate a quantity in calculations where logarithms occur. It is to be desired that these letters d and l be altered to give a different appearance, lest they be confused with other letters of the alphabet that are used to designate quantities. This is what has happened to the letter r, which first was used to indicate a root; the r has been distorted to $\surd$.

120. If y is any function of x, as we have seen, its first differential will have the form $P\omega$. Since $\omega = dx$, we have $dy = P\,dx$. Whatsoever function of x y might be, its differential is expressed by the product of a certain function of x that we call P and the differential of x, that is, dx. Although the differentials of x and y are both infinitely small, and hence equal to zero, still there is a finite ratio between them. That is, $dy : dx = P : 1$. Once we have found the function P, then we know the ratio between the differential dx and the differential dy. Since differential calculus consists in finding differentials, the work involved is not in finding the differentials themselves, which are both equal to zero, but rather in their mutual geometric ratio.

121. Differentials are much easier to find than finite differences. For the finite difference Δy by which a function increases when the variable quantity x increases by ω, it is not sufficient to know P, but we must investigate also the functions $Q, R, S, \ldots$ that enter into the finite difference that we have expressed as

$$P\omega + Q\omega^2 + R\omega^3 + \cdots .$$

For the differential of y we need only to know the function P. For this reason, from our knowledge of the finite difference of any function of x we can easily define its differential. On the other hand, from a function's differential it is not possible to figure out its finite difference. Nevertheless, we shall see (in paragraph 49 of the second part) that from a knowledge of the differentials of all orders it is possible to find the finite difference of any given function. Now, from what we have seen, it is clear that the first differential $dy = P\,dx$ gives the first term of the finite difference, that is, $P\omega$.

122. If the increment ω by which the variable x is considered to be increased happens to be very small, so that in the expression $P\omega + Q\omega^2 + R\omega^3 + \cdots$ the terms $Q\omega^2$ and $R\omega^3$, and even more so the remaining terms, become so small in comparison to $P\omega$ that they can be neglected in computations where rigor is not so important, in this case when we know the differential Pdx we also know approximately the finite difference $P\omega$. Hence, in many cases we can use calculus in applications with no little profit. There are some who judge that differentials are very small increments, but they deny that they are actually equal to zero, and so they say that they are only indefinitely small. This idea presents to others an occasion to blame analysis of the infinite for not obtaining exact, but only approximate, results. This objection has some justification unless we insist that the infinitely small is absolutely equal to zero.

123. Those who are unwilling to admit that the infinitely small becomes nothing, in order that they might seem to meet the objection, compare differentials to the very smallest speck of dust in relation to the whole earth. One is thought not to have given the true bulk of the earth who departs by one speck from the truth. They want such a ratio between a finite quantity and one infinitely small to be as is the ratio between the whole earth and the smallest speck. If there is someone for whom this difference is not sufficiently large, then let the ratio be magnified by even more than a thousand, so that the smallness cannot possibly be observed. However, they are forced to admit that geometrical rigor has been a bit compromised, and to meet this objection they turn to such examples as they may find from geometry or analysis of the infinite; from any agreement between these latter methods they try to draw some good. This argument does not work, since they frequently try to draw the truth from erroneous arguments. In order that an argument avoid this difficulty and even be completely successful, those quantities that we neglect in our calculations must not be just incomprehensibly small, but they must be actually nothing, as we have assumed. In this way geometric rigor suffers absolutely no violence.

124. Let us move on to an explanation of differentials of the second order. These arise from second differences, which were treated in the first chapter, when we let ω become the infinitely small dx. If we suppose that the variable x increases by equal increments, then the second value x^{I} becomes equal to $x + dx$, and the following will be $x^{\mathrm{II}} = x + 2dx$, $x^{\mathrm{III}} = x + 3dx, \ldots$. Since the first differences dx are constant, the second differences vanish, and so the second differential of x, that is, d^2x, is equal to 0. For this reason all of the other differentials of x are equal to 0, namely, $d^3x = 0$, $d^4x = 0$, $d^5x = 0, \ldots$. One could object that since differentials are infinitely small, for that reason alone they are equal to 0, so that there is nothing special about the variable x, whose increments are considered to be equal. However, this vanishing should be interpreted as due not only to the fact that $d^2x, d^3x, \ldots$ are nothing in themselves, but also by reason of the powers of dx, which vanish when compared to dx itself.

125. In order that this may become clearer, let us recall that the second difference of any function y of x can be expressed as $P\omega^2 + Q\omega^3 + R\omega^4 + \cdots$. Hence, if ω should be infinitely small, then the terms $Q\omega^3$, $R\omega^4, \ldots$ vanish when compared with the first term $P\omega^2$, so that with $\omega = dx$, the second differential of y will be equal to $P\,dx^2$, where dx^2 means the square of the differential dx. It follows that although the second differential of y, namely d^2y, by itself is equal to 0, still, since $d^2y = P\,dx^2$, d^2y has a finite ratio to dx^2, that is, as P to 1. However, since $y = x$, we have $P = 0$, $Q = 0$, $R = 0, \ldots$, so that in this case the second differential of x vanishes, even with respect to dx^2, and so do the other higher powers of dx. This is the sense in which we should understand what was stated previously, namely, $d^2x = 0$, $d^3x = 0, \ldots$.

126. Since the second difference is just the difference of the first difference, the second differential, or, as it is frequently called, the differentiodifferential, is the differential of the first differential. Now, since a constant function undergoes no increment or decrement, it has no differences. Strictly speaking, only variable quantities have differentials, but we say that constant quantities have differentials of all orders equal to 0, and hence all powers of dx vanish. Since the differential of dx, that is d^2x, is equal to 0, the differential dx can be thought of as a constant quantity; as long as the differential of any quantity is constant, then that quantity is understood to be taking on equal increments. Here we are taking x to be the quantity whose differential is constant, and thus we estimate the variability of all the functions on which the differentials depend.

127. We let the first differential of y be $p\,dx$. In order to find the second differential we have to find the differential of $p\,dx$. Since dx is a constant and does not change, even though we write $x + dx$ for x, we need only find the differential of the first quantity p. Now let $dp = q\,dx$, since we have seen that the differential of every function of x can be put into this form. From what we have shown for finite differences, we see that the differential of np is equal to $nq\,dx$, where n is a constant quantity. We substitute dx for the constant n, so that the differential of $p\,dx$ is equal to $q\,dx^2$. For this reason, if $dy = p\,dx$ and $dp = q\,dx$, then the second differential $d^2y = q\,dx^2$, and so it is clear, as we indicated before, that the second differential of y has a finite ratio to dx^2.

128. In the first chapter we noticed that the second and higher differences cannot be determined unless the successive values of x are assumed to follow some rule; since this rule is arbitrary, we have decided that the best and easiest rule is that of an arithmetic progression. For the same reason we cannot state anything certain about second differentials unless the first differentials, by which the variable x is thought to increase constantly, follow the stated rule. Hence we suppose that the first differentials of x, namely, dx, dx^{I}, $dx^{\mathrm{II}}, \ldots$, are all equal to each other, so that the second differentials are given by

$$d^2x = dx^{\mathrm{I}} - dx = 0, \qquad d^2x^{\mathrm{I}} = dx^{\mathrm{II}} - dx^{\mathrm{I}} = 0, \qquad \ldots.$$

Since the second differentials, and those of higher order, depend on the order by which the differentials of x are mutually related, and this order is arbitrary, first differentials are not affected by this, and this is the huge difference between the method for finding first differentials and those of higher order.

129. If the successive values of x, namely, x, x^{I}, x^{II}, x^{III}, $x^{\mathrm{IV}}, \ldots$, do not form an arithmetic progression, but follow some other rule, then their first differentials, namely, dx, dx^{I}, $dx^{\mathrm{II}}, \ldots$, will not be equal to each other, and so we do not have $d^2x = 0$. For this reason the second differentials are functions of x with a different form. If the first differential of such a function y is equal to $p\,dx$, to find the second differential it is not enough to multiply the differential of p by dx, but we must also take the differential of dx, that is, d^2x into account. Since the second differential arises from the difference when $p\,dx$ is subtracted from its successor, which we obtain by substituting $x+dx$ for x and $dx+d^2x$ for dx, we suppose that the value of this successor of p has the form $p+q\,dx$, and the successor of $p\,dx$ has the form

$$(p+q\,dx)\left(dx+d^2x\right)=p\,dx+p\,d^2x+q\,dx^2+q\,dx\,d^2x.$$

When we subtract $p\,dx$ from this we have the second differential

$$d^2y=p\,d^2x+q\,dx^2+q\,dx\,d^2x=p\,d^2x+q\,dx^2$$

since $q\,dx\,d^2x$ vanishes when compared to $p\,d^2x$.

130. Although it is simplest and most convenient to have the increments of x equal to each other, nevertheless it is frequently the case that y is not directly a function of x, but a function of some other quantity that is a function of x. Furthermore, frequently it is specified that the first differentials of this other quantity should be equal, but their relation to x may not be clear. In the previous case the second and following differentials of x depend on a relationship that x has with that quantity, and we suppose that the change is by equal increments. In this other case the second and following differentials of x are considered to be unknowns, and we use the symbols d^2x, d^3x, $d^4x,\ldots$.

131. The methods by which these differentiations in the different cases are to be treated we shall discuss at length later. Now we will proceed under the assumption that x increases uniformly, so that the first differentials dx, dx^{I}, $d^{\mathrm{II}},\ldots$ are equal to each other, so that the second and higher differentials are equal to zero. We can state this condition by saying that the differential of x, that is dx, is assumed to be constant. Let y be any function of x; since the function is defined by x and constants, its first, second, third, fourth, and so forth, differentials can be expressed in terms of x and dx. For example, if in y we substitute $x+dx$ for x and subtract the original value of y, there remains the first differential dy. If in this differential we substitute $x+dx$ for x, we obtain dy^{I} and $d^2y=dy^{\mathrm{I}}-dy$. In a similar way, by substituting $x+dx$ for x in d^2y we obtain d^2y^{I} and $d^2y^{\mathrm{I}}-d^2y=d^3y$, and so forth. In all of the calculations dx is always seen as a constant whose differential vanishes.

132. From the definition of y, a function of x, we determine the value of the function p, which when multiplied by dx gives the first differential dy. We can determine p either by the method of finite differences, or by a much more expeditious method that we will discuss later. Given $dy=p\,dx$, the differential of $p\,dx$ gives the second differential d^2y. Hence, if $dp=q\,dx$, since dx is constant, we have $d^2y=q\,dx^2$, as we have already shown. Taking another step, since the differential of the second differential gives the third differential, we let $dq = r\,dx$, so that $d^3y = r\,dx^3$. In like manner, if the differential of this function r is sought, it will be $dr = s\,dx$, from which we obtain the fourth differential $d^4y = s\,dx^4$, and so forth. Provided that we can find the first differential of any function, we can find the differential of any order.

133. In order that we may keep the form of these differentials, and the method of discovery, in mind we present the following table: If y is any function of x,

then	and we let
$dy = p\,dx,$	$dp = q\,dx,$
$d^2y = q\,dx^2,$	$dq = r\,dx,$
$d^3y = r\,dx^3,$	$dr = s\,dx,$
$d^4y = s\,dx^4,$	$ds = t\,dx,$
$d^5y = t\,dx^5,$	$\ldots.$

Since the function p is known from y by differentiation, similarly we find q from p, then r from q, then s, and so forth. We can find differentials of any order, provided only that the differential dx remains constant.

134. Since $p, q, r, s, t, \ldots$ are finite quantities, in particular, functions of x, the first differential of y has a finite ratio to the first differential of x, that is, as p to 1. For this reason, the differentials dx and dy are said to be homogeneous. Then, since d^2y has the finite ratio to dx^2 as q to 1, it follows that d^2y and dx^2 are homogeneous. Similarly, d^3y and dx^3 as well as d^4y and dx^4 are homogeneous, and so forth. Hence, just as first differentials are mutually homogeneous, that is, they have a finite ratio, so second differentials with the squares of first differentials, third differentials with cubes of first differentials, and so forth, are homogeneous. In general, the differential of y of the nth order, expressed as d^ny, is homogeneous with dx^n, that is, with the nth power of dx.

135. Since in comparison with dx all of its powers greater than 1 vanish, so also in comparison with dy all of the powers dx^2, dx^3, $dx^4, \ldots$ vanish, as well as the differentials of higher orders that have finite ratios with these, that is, d^2y, d^3y, $d^4y, \ldots$. In a similar way, in comparison with d^2y, since this is homogeneous with dx^2, all powers of dx that are greater than the second, dx^3, $dx^4, \ldots$, will vanish. Along with these will vanish d^3y, $d^4y, \ldots$. Furthermore, compared to d^3y, we have dx^4, d^4y, dx^5, $d^5y, \ldots$ all vanishing. Hence, given expressions involving differentials of this kind, it is easy to decide whether or not they are homogeneous. We have only to consider the differentials, since the finite parts do not disturb homogeneity.

For differentials of the second and higher order, consider the powers of dx; if the numbers are the same, the expressions are homogeneous.

136. Thus it is clear that the expressions $P\, d^2y^2$ and $Q\, dy\, d^3y$ are mutually homogeneous. For d^2y^2 is the square of d^2y, and since d^2y is homogeneous with dx^2, it follows that d^2y^2 is homogeneous with dx^4. Thus, since dy is homogeneous with dx and d^3y is homogeneous with dx^3, we have that the product $dy\, d^3y$ is homogeneous with dx^4. From this it follows that $P\, d^2y^2$ and $Q\, dy\, d^3y$ are mutually homogeneous, and so their ratio is finite. Similarly, we gather that the expressions

$$\frac{P\, d^3y^2}{dx\, d^2y} \quad \text{and} \quad \frac{Q\, d^5y}{dy^2}$$

are homogeneous. If we substitute for dy, d^2y, d^3y, and d^5y the powers of dx that are homogeneous with them, namely, dx, dx^2, dx^3, and dx^5, we obtain the expressions $P\, dx^3$ and $Q\, dx^3$, which are mutually homogeneous.

137. If after reduction the proposed expressions do not contain the same powers of dx, then the expressions are not homogeneous, nor is their ratio finite. In this case one will be either infinitely greater or infinitely less than the other, and so one will vanish with respect to the other. Thus $P\, d^3y/dx^2$ to $Q\, d^2y^2/dy$ has a ratio infinitely large. The former reduces to $P\, dx$ and the latter to $Q\, dx^3$. It follows that the latter will vanish when compared to the former. For this reason, if in some calculation the sum of these two terms occurs

$$\frac{P\, d^3y}{dx^2} + \frac{Q\, d^2y^2}{dy},$$

the second term, compared to the first, can safely be eliminated, and only the first term $P\, d^3y/dx^2$ is kept in the calculation. There is a perfect ratio of equality between the expressions

$$\frac{P\, d^3y}{dx^2} + \frac{Q\, d^2y^2}{dy} \quad \text{and} \quad \frac{P\, d^3y}{dx^2},$$

since when we express the ratio, we obtain

$$1 + \frac{Q\, dx^2 d^2y^2}{P\, dy\, d^3y} = 1, \qquad \text{because} \qquad \frac{Q\, dx^2 d^2y^2}{P\, dy\, d^3y} = 0.$$

In this way differential expressions can sometimes be wonderfully reduced.

138. In differential calculus rules are given by means of which the first differential of a given quantity can be found. Since second differentials are obtained by differentiating first differentials, third differentials by the same operation on seconds, and so forth, the next one from the one just found, differential calculus contains a method for finding all differentials of each order. From the word *differential*, which denotes an infinitely small difference, we derive other names that have come into common usage. Thus we have the word *differentiate*, which means *to find a differential.* A quantity is said to be *differentiated* when its differential is found. *Differentiation* denotes the operation by which differentials are found. Hence differential calculus is also called the method of *differentiating*, since it contains a way of finding differentials.

139. Just as in differential calculus the differential of any quantity is investigated, so there is a kind of calculus that consists in finding a quantity whose differential is one that is already given, and this is called *integral calculus.* If any differential is given, that quantity whose differential is the proposed quantity is called its *integral.* The reason for this name is as follows: Since a differential can be thought of as an infinitely small part by which a quantity increases, that quantity with respect to which this is a part can be thought of as a whole, that is, integral, and for this reason is called an integral. Thus, since dy is the differential of y, y, in turn, is the integral of dy. Since d^2y is the differential of dy, dy is the integral of d^2y. Likewise, d^2y is the integral of d^3y, and d^3y is the integral of d^4y, and so forth. It follows that any differentiation, from an inverse point of view, is also an example of integration.

140. The origin and nature of both integrals and differentials can most clearly be explained from the theory of finite differences, which has been discussed in the first chapter. After it was shown how the difference of any quantity should be found, going in reverse, we also showed how, from a given difference, a quantity can be found whose difference is the one proposed. We called that quantity, with respect to its difference, the sum of the difference. Just as when we proceed to the infinitely small, differences become differentials, so the sums, which there were called just that, now receive the name of integral. For this reason integrals are sometimes called sums. The English call differentials by the name fluxions, and integrals are called by them fluents. Their mode of speaking about finding the fluent of a given fluxion is the same as ours when we speak of finding the integral of a given differential.

141. Just as we use the symbol d for a differential, so we use the symbol $\int$ to indicate an integral. Hence if this is placed before a differential, we are indicating that quantity whose differential is the one given. Thus, if the differential of y is $p\,dx$, that is, $dy = p\,dx$, then y is the integral of $p\,dx$. This is expressed as follows: $y = \int p\,dx$, since $y = \int dy$. Hence, the integral of $p\,dx$, symbolized by $\int p\,dx$, is that quantity whose differential is $p\,dx$. In a similar way if $d^2y = q\,dx$, where $dp = q\,dx$, then the integral of d^2y is dy, which is equal to $p\,dx$. Since $p = \int q\,dx$, we have $dy = dx \int q\,dx$, and hence $y = \int dx \int q\,dx$. If in addition, $dq = r\,dx$, then $q = \int r\,dx$ and

$dp = dx \int r\,dx$, so that if we place the symbol $\int$ before both sides, we have $p = \int dx \int r\,dx$. Finally, we have $dy = dx \int dx \int r\,dx$ and so $y = \int dx \int dx \int r\,dx$.

142. Since the differential dy is an infinitely small quantity, its integral y is a finite quantity. In like manner the second differential d^2y is infinitely less that its integral dy. It should be clear that a differential will vanish in the presence of its integral. In order that this relation be better understood, the infinitely small can be categorized by orders. First differentials are said to be infinitely small of the first order; the infinitely small of the second order consist of differentials of the second order, which are homogeneous with dx^2. Similarly, the infinitely small that are homogeneous with dx^3 are said to be of the third order, and these include all differentials of the third order, and so forth. Hence, just as the infinitely small of the first order vanish in the presence of finite quantities, so the infinitely small of the second order vanishes in the presence of the infinitely small of the first order. In general, the infinitely small of any higher order vanishes in the presence of an infinitely small of a lower order.

143. Once the orders of the infinitely small have been established, so that the differential of a finite quantity is infinitely small of the first order, and so forth, conversely, the integral of an infinitely small of the first order is a finite quantity. The integral of an infinitely small of the second order is an infinitely small of the first order, and so forth. Hence if a given differential is infinitely small of order n, then its integral will be infinitely small of order $n-1$. Thus, just as differentiating increases the order of the infinitely small, so integrating lowers the order until we come to a finite quantity. If we wished to integrate again finite quantities, then according to this law we obtain quantities infinitely large. From the integration of these we get quantities infinitely greater still. Proceeding in this way we obtain orders of infinity such that each one is infinitely greater than its predecessor.

144. It remains to give something of a warning about the use of symbols in this chapter, lest there still be any ambiguity. First of all, the symbol for differentiation, d, operates on only the letter that comes immediately after it. Thus, $dx\,y$ does not mean the differential of the product xy, but rather the product of y and the differential of x. In order to minimize the confusion we ordinarily would write this with the y preceding the symbol d, as $y\,dx$, by which we indicate the product of y and dx. If y happens to be a quantity preceded by a symbol indicating either a root $\sqrt{\ }$ or a logarithm, then we usually place that after the differential. For instance, $dx\sqrt{a^2 - x^2}$ signifies the product of the finite quantity $\sqrt{a^2 - x^2}$ and the differential dx. In like manner, $dx \ln(1+x)$ is the product of the logarithm of the quantity $1+x$ and dx. For the same reason $d^2y\sqrt{x}$ expresses the product of the second differential d^2y and the finite quantity $\sqrt{x}$.

145. The symbol d does not affect only the letter immediately following it, but also the exponent on that letter if it has one. Thus, dx^2 does not express the differential of x^2, but the square of the differential of x, so that the exponent 2 refers not to x but to dx. We could write this as $dx\,dx$ in the same way as we would write the product of two differentials dx and dy as $dx\,dy$. The previous method, dx^2, has the advantage of being both briefer and more usual. Especially if it is a question of higher powers of dx, the method of repeating so many times tends to be too long. Thus, dx^3 denotes the cube of dx; we observe the same reasoning with regard to differentials of higher order. For example, d^2y^4 denotes the fourth power of the second-order differential d^2y, and $d^3y^2\sqrt{x}$ symbolizes the product of the square of the differential of the third order of y and $\sqrt{x}$. If it were the product with the rational quantity x then we would write it as $x\,d^3y^2$.

146. If we want the symbol d to affect more than the next letter, we need a special way of indicating that. In this case we will use parentheses to include the expression whose differential we need to express. Then $d\left(x^2+y^2\right)$ means the differential of the quantity x^2+y^2. It is true enough that if we want to designate the differential of a power of such an expression, then ambiguity can hardly be avoided. If we write $d\left(x^2+y^2\right)^2$, this could be understood to mean the square of $d\left(x^2+y^2\right)$. On the other hand, we can avoid this difficulty with the use of a dot, so that $d.\left(x^2+y^2\right)^2$ means the differential of $\left(x^2+y^2\right)^2$. If the dot is missing, then $d\left(x^2+y^2\right)^2$ indicates the square of $d\left(x^2+y^2\right)$. The dot conveniently indicates that the symbol d applies to the whole expression after the dot. Thus, $d.x\,dy$ expresses the differential of $x\,dy$, and $d^3.x\,dy\sqrt{a^2+x^2}$ is the third-order differential of the expression $x\,dy\sqrt{a^2+x^2}$, which is the product of the finite quantities x and $\sqrt{a^2+x^2}$ and the differential dy.

147. On the one hand, the symbol for differentiation d affects only the quantity immediately following it, unless a dot intervenes and extends its influence to the whole following expression; on the other hand, the integral sign $\int$ always extends to the whole expression that follows. Thus, $\int y\,dx\left(a^2-x^2\right)^n$ denotes the integral of, or the quantity whose differential is, $y\,dx\left(a^2-x^2\right)^n$. The expression $\int x\,dx\int dx\,\ln x$ denotes the quantity whose differential is $x\,dx\int dx\,\ln x$. Hence, if we wish to express the product of two integrals, for instance $\int y\,dx$ and $\int z\,dx$, it would be wrong to write $\int y\,dx\int z\,dx$. This would be understood as the integral of $y\,dx\int z\,dx$. For this reason we again use a dot to remove any ambiguity, so that $\int y\,dx\cdot\int z\,dx$ signifies the product of the integrals $\int y\,dx$ and $\int z\,dx$.

148. Now, analysis of the infinite is concerned with the discovery of both differentials and integrals, and for this reason it is divided into two principal parts, one of which is called differential calculus, and the other is integral calculus. In the former, rules are given for finding differentials of any quantity. In the latter a way of investigating integrals of given differentials is shown. In both parts there are indications of the best applications to both analysis itself and higher geometry. For this reason even this first part of analysis has already grown so that to cover it requires no small book. In the integral calculus both new methods of integration are being discovered every day, as well as the revelation of new aids for the solution of different kinds of problems. Due to the new discoveries that are continuously being made, we could never exhaust, much less describe and explain perfectly, all of this. Nevertheless I will make every effort in these books to make sure that either everything that has so far been discovered shall be presented, or at least the methods by which they can be deduced are explained.

149. It is common to give other parts of analysis of the infinite. Besides differential and integral calculus, one sometimes finds differentio-differential calculus and exponential calculus. In differentio-differential calculus the methods of finding second and higher differentials are usually discussed. Since the method of finding differentials of any order will be discussed in this differential calculus, this subdivision, which seems to be based more on the importance of its discovery rather than the thing itself, we will omit. The illustrious Johann Bernoulli, to whom we are eternally grateful for innumerable and great discoveries in analysis of the infinite, extended the methods of differentiating and integrating to exponential quantities by means of exponential calculus. Since I plan to treat in both parts of calculus not only algebraic but also transcendental quantities, this special part has become superfluous and outside our plan.

150. I have decided to treat differential calculus first. I will explain the method by which not only first differentials but also second and higher differentials of variable quantities can be expeditiously found. I will begin by considering algebraic quantities, whether they be explicitly given or implicitly by equations. Then I will extend the discovery of differentials to nonalgebraic quantities, at least to those which can be known without the aid of integral calculus. Quantities of this kind are logarithms and exponential quantities, as well as arcs of circles and in turn sines and tangents of circular arcs. Finally, we will teach how to differentiate compositions and mixtures of all of these quantities. In short, this first part of differential calculus will be concerned with differentiating.

151. The second part will be dedicated to the explanation of the applications of the method of differentiating to both analysis and higher geometry. Many nice things spill over into ordinary algebra: finding roots of equations, discussing and summing series, discovering maxima and minima, defining and discovering values of expressions that in some cases seem to defy determination. Higher geometry has received its greatest development from differential calculus. By its means tangents to curves and their curvature can be defined with marvelous facility. Many other problems concerned with either reflex or refracted radii of curves can be solved. Although a long treatise could be devoted to all of this, I will endeavor, as far as possible, to give a brief and clear account.

5

On the Differentiation of Algebraic Functions of One Variable

152. Since the differential of the variable x is equal to dx, when x is incremented, x becomes equal to $x + dx$. Hence, if y is some function of x, and if we substitute $x + dx$ for x, we obtain y^{I}. The difference $y^{\mathrm{I}} - y$ gives the differential of y. Now if we let $y = x^n$, then

$$y^{\mathrm{I}} = (x + dx)^n = x^n + nx^{n-1}dx + \frac{n(n-1)}{1 \cdot 2}x^{n-2}dx^2 + \cdots,$$

and so

$$dy = y^{\mathrm{I}} - y = nx^{n-1}dx + \frac{n(n-1)}{1 \cdot 2}x^{n-2}dx^2 + \cdots.$$

In this expression the second term and all succeeding terms vanish in the presence of the first term. Hence, $nx^{n-1}dx$ is the differential of x^n, or

$$d.x^n = nx^{n-1}dx.$$

It follows that if a is a number or constant quantity, then we also have $d.ax^n = nax^{n-1}dx$. Therefore, the differential of any power of x is found by multiplying that power by the exponent, dividing by x, and multiplying the result by dx. This rule can easily be memorized.

153. Once we know the first differential of x^n, it is easy to find its second differential, provided that we assume that the differential dx remains constant. Since in the differential $nx^{n-1}dx$ the factor $n\,dx$ is constant, the differential of the other factor x^{n-1} must be taken, which will be

$(n-1)\,x^{n-2}dx$. When this is multiplied by $n\,dx$, we have the second differential

$$d^2.x^n = n\,(n-1)\,x^{n-2}dx^2.$$

In a similar way, if the differential of x^{n-2}, which is equal to $(n-2)\,x^{n-3}dx$, is multiplied by $n\,(n-1)\,dx^2$, we have the third differential

$$d^3.x^n = n\,(n-1)\,(n-2)\,x^{n-3}dx^3.$$

Furthermore, the fourth differential will be

$$d^4.x^n = n\,(n-1)\,(n-2)\,(n-3)\,x^{n-4}dx^4,$$

and the fifth differential is

$$d^5.x^n = n\,(n-1)\,(n-2)\,(n-3)\,(n-4)\,x^{n-5}dx^5.$$

The form of the following differentials is easily understood.

154. As long as n is a positive integer, eventually the higher differentials will vanish; these are equal to 0, because differentials of higher powers of dx vanish. We should note a few of the simpler cases:

$$\begin{array}{llll}
d.x = dx, & d^2.x = 0, & d^3.x = 0, & \ldots, \\
d.x^2 = 2x\,dx, & d^2.x^2 = 2dx^2, & d^3.x^2 = 0, & d^4.x^2 = 0, \ \ldots, \\
d.x^3 = 3x^2dx, & d^2.x^3 = 6x\,dx^2, & d^3.x^3 = 6dx^3, & d^4.x^3 = 0, \ \ldots, \\
d.x^4 = 4x^3dx, & d^2.x^4 = 12x^2dx^2, & d^3.x^4 = 24x\,dx^3, & d^4.x^4 = 24dx^4, \\
d^5.x^4 = 0 & \ldots, & & \\
d.x^5 = 5x^4dx, & d^2.x^5 = 20x^3dx^2, & d^3.x^5 = 60x^2dx^3, & \\
d^4.x^5 = 120x\,dx^4, & & & \\
d^5.x^5 = 120dx^5, & d^6.x^5 = 0, & \ldots. &
\end{array}$$

It is clear that if n is a positive integer, then the differential of order n of x^n will be a constant, that is, it will be equal to $1 \cdot 2 \cdot 3 \cdots n\,dx^n$. The result is that all differentials of higher order will be equal to 0.

155. If n is a negative integer, differentials of x with such negative powers can be taken, such as

$$\frac{1}{x}, \quad \frac{1}{x^2}, \quad \frac{1}{x^3}, \quad \ldots,$$

since

$$\frac{1}{x} = x^{-1}, \qquad \frac{1}{x^2} = x^{-2},$$

and generally,

$$\frac{1}{x^m} = x^{-m}.$$

If we substitute in the previous formula $-m$ for n, we have the first differential of $1/x^m$ equal to

$$\frac{-m\,dx}{x^{m+1}};$$

the second differential is equal to

$$\frac{m\,(m+1)\,dx^2}{x^{m+2}};$$

the third differential is equal to

$$\frac{-m\,(m+1)\,(m+2)\,dx^3}{x^{m+3}};$$

and so forth. The following simpler cases deserve to be noted:

$$d.\frac{1}{x} = \frac{-dx}{x^2}, \qquad d^2.\frac{1}{x} = \frac{2dx^2}{x^3}, \qquad d^3.\frac{1}{x} = \frac{-6dx^3}{x^4}, \qquad \ldots,$$

$$d.\frac{1}{x^2} = \frac{-2dx}{x^3}, \qquad d^2.\frac{1}{x^2} = \frac{6dx^2}{x^4}, \qquad d^3.\frac{1}{x^2} = \frac{-24dx^3}{x^5}, \qquad \ldots,$$

$$d.\frac{1}{x^3} = \frac{-3dx}{x^4}, \qquad d^2.\frac{1}{x^3} = \frac{12dx^2}{x^5}, \qquad d^3.\frac{1}{x^3} = \frac{-60dx^3}{x^6}, \qquad \ldots,$$

$$d.\frac{1}{x^4} = \frac{-4dx}{x^5}, \qquad d^2.\frac{1}{x^4} = \frac{20dx^2}{x^6}, \qquad d^3.\frac{1}{x^4} = \frac{-120dx^3}{x^7}, \qquad \ldots,$$

$$d.\frac{1}{x^5} = \frac{-5dx}{x^6}, \qquad d^2.\frac{1}{x^5} = \frac{30dx^2}{x^7}, \qquad d^3.\frac{1}{x^5} = \frac{-210dx^3}{x^8}, \qquad \ldots,$$

and so forth.

156. Then if we let n be a fraction, we obtain differentials of irrational expressions. If $n = \mu/\nu$, then the first differential of $x^{\mu/\nu}$, that is $\sqrt[\nu]{x^\mu}$, is equal to

$$\frac{\mu}{\nu} x^{(\mu-\nu)/\nu} dx = \frac{\mu}{\nu} dx \sqrt[\nu]{x^{\mu-\nu}}.$$

The second differential is equal to

$$\frac{\mu\,(\mu-\nu)}{\nu^2} x^{(\mu-2\nu)/\nu} dx^2 = \frac{\mu\,(\mu-\nu)}{\nu^2} dx^2 \sqrt[\nu]{x^{\mu-2\nu}},$$

and so forth. Hence we have

$$d.\sqrt{x} = \frac{dx}{2\sqrt{x}}, \quad d^2.\sqrt{x} = \frac{-dx^2}{4x\sqrt{x}}, \quad d^3.\sqrt{x} = \frac{1\cdot 3dx^3}{8x^2\sqrt{x}}, \quad \ldots,$$

$$d.\sqrt[3]{x} = \frac{dx}{3\sqrt[3]{x^2}}, \quad d^2.\sqrt[3]{x} = \frac{-2dx^2}{9x\sqrt[3]{x^2}}, \quad d^3.\sqrt[3]{x} = \frac{2\cdot 5dx^3}{27x^2\sqrt[3]{x^2}}, \quad \ldots,$$

$$d.\sqrt[4]{x} = \frac{dx}{4\sqrt[4]{x^3}}, \quad d^2.\sqrt[4]{x} = \frac{-3dx^2}{16x\sqrt[4]{x^3}}, \quad d^3.\sqrt[4]{x} = \frac{3\cdot 7dx^3}{64x^2\sqrt[4]{x^3}}, \quad \ldots.$$

If we inspect these expressions a bit, we can easily find the differentials, even without putting the expression into exponential form.

157. If μ is not 1, but some other integer, whether positive or negative, the differentials can be defined just as easily. Since the second- and higher-order differentials are defined from the first, using the same law of exponents, we put down a few of the simpler examples of only first differentials.

$$d.x\sqrt{x} = \frac{3}{2}dx\sqrt{x}, \quad d.x^2\sqrt{x} = \frac{5}{2}x\,dx\sqrt{x}, \quad d.x^3\sqrt{x} = \frac{7}{2}x^2dx\sqrt{x}, \quad \ldots,$$

$$d.\frac{1}{\sqrt{x}} = \frac{-dx}{2x\sqrt{x}}, \quad d.\frac{1}{x\sqrt{x}} = \frac{-3dx}{2x^2\sqrt{x}}, \quad d.\frac{1}{x^2\sqrt{x}} = \frac{-5dx}{2x^3\sqrt{x}}, \quad \ldots,$$

$$d.\sqrt[3]{x^2} = \frac{2dx}{3\sqrt[3]{x}}, \quad d.x\sqrt[3]{x} = \frac{4}{3}dx\sqrt[3]{x}, \quad d.x\sqrt[3]{x^2} = \frac{5}{3}dx\sqrt[3]{x^2},$$

$$d.x^2\sqrt[3]{x} = \frac{7}{3}x\,dx\sqrt[3]{x}, \quad d.x^2\sqrt[3]{x} = \frac{8}{3}x\,dx\sqrt[3]{x}, \quad \ldots,$$

$$d.\frac{1}{\sqrt[3]{x}} = \frac{-dx}{3x\sqrt[3]{x}}, \quad d.\frac{1}{\sqrt[3]{x^2}} = \frac{-2dx}{3x\sqrt[3]{x^2}}, \quad d.\frac{1}{x\sqrt[3]{x}} = \frac{-4dx}{3x^2\sqrt[3]{x}},$$

$$d.\frac{1}{x\sqrt[3]{x^2}} = \frac{-5dx}{3x^2\sqrt[3]{x^2}}, \quad d.\frac{1}{x^2\sqrt[3]{x}} = \frac{-7dx}{3x^3\sqrt[3]{x}}, \quad \ldots.$$

158. From functions of this kind we can find the differentials of all rational algebraic functions, since each of their terms is a power of x, which we know how to differentiate. Suppose we have a quantity of the form

$$p + q + r + s + \cdots.$$

When we substitute $x + dx$ for x we obtain

$$p + dp + q + dq + r + dr + s + ds + \cdots,$$

so that its differential is equal to

$$dp + dq + dr + ds + \cdots .$$

Hence, if we can give the differential of each quantity p, q, r, s, then we know the differential of the sum. Furthermore, since the differential of a multiple of p is the same multiple of dp, we have $d.ap = a\,dp$, and the differential of $ap + bq + cr$ is equal to $a\,dp + b\,dq + c\,dr$. Finally, since the differentials of constants are zero, the differential of $ap+bq+cr+f$ is equal to $a\,dp + b\,dq + c\,dr$.

159. In polynomial functions, since each term is either a constant or power of x, differentiation according to the given rule is easily carried out. Thus we have

$$\begin{aligned}
d\,(a+x) &= dx,\\
d\,(a+bx) &= b\,dx,\\
d\left(a+x^2\right) &= 2x\,dx,\\
d\left(a^2-x^2\right) &= -2x\,dx,\\
d\left(a+bx+cx^2\right) &= b\,dx + 2cx\,dx,\\
d\left(a+bx+cx^2+ex^3\right) &= b\,dx + 2cx\,dx + 3ex^2dx,\\
d\left(a+bx+cx^2+ex^3+fx^4\right) &= b\,dx + 2cx\,dx + 3ex^2dx + 4fx^3dx.
\end{aligned}$$

If the exponents are indefinite, then

$$\begin{aligned}
d\,(1-x^n) &= -nx^{n-1}dx,\\
d\,(1+x^m) &= mx^{m-1}dx,\\
d\,(a+bx^m+cx^n) &= mbx^{m-1}dx + ncx^{n-1}dx.
\end{aligned}$$

160. Since the degree of a polynomial is given by the term with the highest power of x, it is clear that if differentials of such functions are continually taken, the differential will eventually become constant and then vanish, provided that we assume that dx is constant. Thus the first differential of a first degree polynomial $a+bx$, $b\,dx$, is constant, and the second and higher differentials vanish. Let $a + bx + cx^2 = y$ be a second-degree polynomial. Then

$$dy = b\,dx + 2cx\,dx, \qquad d^2y = 2c\,dx^2, \qquad d^3y = 0.$$

Likewise, if $a + bx + cx^2 + ex^3 = y$ is a third-degree polynomial, then

$$\begin{aligned} dy &= b\,dx + 2cx\,dx + 3ex^2\,dx \\ d^2y &= 2c\,dx^2 + 6ex\,dx^2, \\ d^3y &= 6e\,dx^3, \\ d^4y &= 0. \end{aligned}$$

In general, if the function is of degree n, then its differential of order n will be constant, and higher-order differentials will all vanish.

161. Nor is there any difficulty with differentiation if among the powers of x that make up a function we have negative or fractional exponents. Thus

I. If

$$y = a + b\sqrt{x} - \frac{c}{x},$$

then

$$dy = \frac{b\,dx}{2\sqrt{x}} + \frac{c\,dx}{x^2}.$$

II. If

$$y = \frac{a}{\sqrt{x}} + b + c\sqrt{x} - ex,$$

then

$$dy = \frac{-a\,dx}{2x\sqrt{x}} + \frac{c\,dx}{2\sqrt{x}} - e\,dx$$

and

$$d^2y = \frac{3a\,dx^2}{4x^2\sqrt{x}} - \frac{c\,dx^2}{4x\sqrt{x}}.$$

III. If

$$y = a + \frac{b}{\sqrt[3]{x^2}} - \frac{c}{x\sqrt[3]{x}} + \frac{f}{x^2},$$

then

$$dy = \frac{-2b\,dx}{3x\sqrt[3]{x^2}} + \frac{4c\,dx}{3x^2\sqrt[3]{x}} - \frac{2f\,dx}{x^3}$$

and

$$d^2y = \frac{10b\,dx^2}{9x^2\sqrt[3]{x^2}} - \frac{28c\,dx^2}{9x^3\sqrt[3]{x}} + \frac{6f\,dx^2}{x^4}.$$

Further examples of this kind are easily treated according to the given laws.

162. If the quantity proposed for differentiation is the power of some function whose differential we can find, then the preceding rules are sufficient to find the first differential. Let p be any function of x that is raised to some power and whose differential is dp. Then the first differential of p^n is equal to $np^{n-1}dp$. From this we obtain the following.

I. If $y = (a+x)^n$, then

$$dy = n\,(a+x)^{n-1}\,dx.$$

II. If $y = \left(a^2 - x^2\right)^2$, then

$$dy = -4x\,dx\left(a^2 - x^2\right).$$

III. If $y = \dfrac{1}{a^2+x^2} = \left(a^2+x^2\right)^{-1}$, then

$$dy = \frac{-2x\,dx}{\left(a^2+x^2\right)^2}.$$

IV. If $y = \sqrt{a+bx+cx^2}$, then

$$dy = \frac{b\,dx + 2cx\,dx}{2\sqrt{a+bx+cx^2}}.$$

V. If $y = \sqrt[3]{\left(a^4-x^4\right)^2} = \left(a^4-x^4\right)^{2/3}$, then

$$dy = -\frac{8}{3}x^3dx\left(a^4-x^4\right)^{-\frac{1}{3}} = \frac{-8x^3dx}{3\sqrt[3]{a^4-x^4}}.$$

VI. If $y = \dfrac{1}{\sqrt{1-x^2}} = \left(1-x^2\right)^{-\frac{1}{2}}$, then

$$dy = x\,dx\left(1-x^2\right)^{-\frac{3}{2}} = \frac{x\,dx}{\left(1-x^2\right)\sqrt{1-x^2}}.$$

VII. If $y = \sqrt[3]{a+\sqrt{bx}+x}$, then

$$dy = \frac{\left(dx\sqrt{b}\right)/\left(2\sqrt{x}\right)+dx}{3\sqrt[3]{\left(a+\sqrt{bx}+x\right)^2}} = \frac{dx\sqrt{b}+2dx\sqrt{x}}{6\sqrt{x}\sqrt[3]{\left(a+\sqrt{bx}+x\right)^2}}.$$

VIII. If $y = \frac{1}{x + \sqrt{a^2 - x^2}}$, then since

$$d.\sqrt{a^2 - x^2} = \frac{-x\,dx}{\sqrt{a^2 - x^2}},$$

we have

$$dy = \frac{-dx + (x\,dx) \big/ \left(\sqrt{a^2 - x^2}\right)}{\left(x + \sqrt{a^2 - x^2}\right)^2} = \frac{x\,dx - dx\sqrt{a^2 - x^2}}{\left(x + \sqrt{a^2 - x^2}\right)^2 \sqrt{a^2 - x^2}},$$

or

$$dy = \frac{dx\left(x - \sqrt{a^2 - x^2}\right)^3}{\left(2x^2 - a^2\right)^2 \sqrt{a^2 - x^2}}.$$

IX. If $y = \sqrt[4]{\left(1 - \frac{1}{\sqrt{x}} + \sqrt[3]{(1 - x^2)^2}\right)^3}$, we let

$$\frac{1}{\sqrt{x}} = p \qquad \text{and} \qquad \sqrt[3]{(1 - x^2)^2} = q;$$

since $y = \sqrt[4]{(1 - p + q)^3}$, we have

$$dy = \frac{-3dp + 3dq}{4\sqrt[4]{1 - p + q}}.$$

From previous work we have

$$dp = \frac{-dx}{2x\sqrt{x}} \qquad \text{and} \qquad dq = \frac{-4x\,dx}{3\sqrt[3]{1 - x^2}}.$$

When these results are substituted, we have

$$dy = \frac{(3dx) \big/ (2x\sqrt{x}) - (4x\,dx) \big/ \sqrt[3]{1 - x^2}}{4\sqrt[4]{1 - \frac{1}{\sqrt{x}} + \sqrt[3]{(1 - x^2)^2}}}.$$

In a similar way, by substituting individual letters for terms to be composed, we can easily find the differentials of this kind of function.

163. If the quantity that is to be differentiated is the product of two or more functions of x whose differentials are known, the most convenient method for finding the differential is as follows. Let p and q be functions of x with differentials dp and dq already known. When we substitute $x + dx$ for

x, p becomes $p + dp$, and q becomes $q + dq$. The product pq is transformed into

$$(p + dp)(q + dq) = pq + p\,dq + q\,dp + dp\,dq.$$

Hence, the differential of the product pq is equal to $p\,dq + q\,dp + dp\,dq$. Since $p\,dq$ and $q\,dp$ are infinitely small of the first order, while $dp\,dq$ is of the second order, the last term vanishes, with the result that

$$d.pq = p\,dq + q\,dp.$$

It follows that the differential of the product pq consists of two members, each of which is one factor multiplied by the differential of the other. From this we easily deduce the differential of the triple product pqr. If we let $qr = z$, then $pqr = pz$ and $d.pqr = p\,dz + z\,dp$. Since $z = qr$, we have $dz = q\,dr + r\,dq$, and after substituting for z and dz, we have

$$d.pqr = pq\,dr + pr\,dq + qr\,dp.$$

In a similar way, if the quantity to be differentiated is a fourfold product, then we have

$$d.pqrs = pqr\,ds + pqs\,dr + prs\,dq + qrs\,dp.$$

From this it should be easily seen what the differential of a product of many factors will be.

I. If $y = (a + x)(b - x)$, then

$$dy = -dx\,(a + x) + dx\,(b - x) = -a\,dx + b\,dx - 2x\,dx.$$

This same differential can be found by expanding the expression to $y = ab - ax + bx - x^2$, so that by the previous rule,

$$dy = -a\,dx + b\,dx - 2x\,dx.$$

II. If $y = \dfrac{1}{x}\sqrt{a^2 - x^2}$, we let

$$\frac{1}{x} = p \qquad \text{and} \qquad \sqrt{a^2 - x^2} = q,$$

but since

$$dp = \frac{-dx}{x^2} \qquad \text{and} \qquad dq = \frac{-x\,dx}{\sqrt{a^2 - x^2}},$$

we have

$$dy = p\,dq + q\,dp = \frac{-dx}{\sqrt{a^2 - x^2}} - \frac{dx}{x^2}\sqrt{a^2 - x^2}.$$

If we take a common denominator, we have

$$\frac{-x^2dx - a^2dx + x^2dx}{x^2\sqrt{a^2 - x^2}} = \frac{-a^2dx}{x^2\sqrt{a^2 - x^2}}.$$

Hence the desired differential is

$$dy = \frac{-a^2dx}{x^2\sqrt{a^2 - x^2}}.$$

III. If $y = \dfrac{x^2}{\sqrt{a^4 + x^4}}$, we let

$$x^2 = p \qquad \text{and} \qquad \frac{1}{\sqrt{a^4 + x^4}} = q.$$

We find that

$$dp = 2x\,dx \qquad \text{and} \qquad dq = \frac{-2x^3dx}{(a^4 + x^4)^{3/2}},$$

so that

$$p\,dq + q\,dp = \frac{-2x^5dx}{(a^4 + x^4)^{3/2}} + \frac{2x\,dx}{\sqrt{a^4 + x^4}} = \frac{2a^4x\,dx}{(a^4 + x^4)^{3/2}}.$$

It follows that the desired differential is

$$dy = \frac{2a^4x\,dx}{(a^4 + x^4)\sqrt{a^4 + x^4}}.$$

IV. If $y = \dfrac{x}{x + \sqrt{1 + x^2}}$, we let

$$x = p \qquad \text{and} \qquad \frac{1}{x + \sqrt{1 + x^2}} = q.$$

Since

$$dp = dx$$

and

$$dq = \frac{-dx - (x\,dx)/\sqrt{1 + x^2}}{\left(x + \sqrt{1 + x^2}\right)^2} = \frac{-dx\left(x + \sqrt{1 + x^2}\right)}{\left(x + \sqrt{1 + x^2}\right)^2\sqrt{1 + x^2}}$$

$$= \frac{-dx}{\left(x + \sqrt{1 + x^2}\right)\sqrt{1 + x^2}},$$

we have

$$p\,dq + q\,dp = \frac{-x\,dx}{\left(x+\sqrt{1+x^2}\right)\sqrt{1+x^2}} + \frac{dx}{x+\sqrt{1+x^2}}$$

$$= \frac{dx\left(\sqrt{1+x^2}-x\right)}{\left(x+\sqrt{1+x^2}\right)\sqrt{1+x^2}}.$$

Therefore, the desired differential is

$$dy = \frac{dx\left(\sqrt{1+x^2}-x\right)}{\left(x+\sqrt{1+x^2}\right)\sqrt{1+x^2}}.$$

If we multiply both numerator and denominator of this fraction by $\sqrt{1+x^2}-x$, we have

$$dy = \frac{dx\left(1+2x^2-2x\sqrt{1+x^2}\right)}{\sqrt{1+x^2}} = \frac{dx+2x^2dx}{\sqrt{1+x^2}} - 2x\,dx.$$

The same differential can be more easily obtained. Since

$$y = \frac{x}{x+\sqrt{1+x^2}},$$

if we multiply both numerator and denominator by $\sqrt{1+x^2}-x$, we have

$$y = x\sqrt{1+x^2} - x^2 = \sqrt{x^2+x^4} - x^2.$$

By the previous rule we have

$$dy = \frac{x\,dx+2x^3dx}{\sqrt{x^2+x^4}} - 2x\,dx = \frac{dx+2x^2dx}{\sqrt{1+x^2}} - 2x\,dx.$$

V. If $y = (a+x)(b-x)(x-c)$, then

$$dy = (a+x)(b-x)\,dx - (a+x)(x-c)\,dx + (b-x)(x-c)\,dx.$$

VI. If $y = x\left(a^2+x^2\right)\sqrt{a^2-x^2}$, because of the three factors we have

$$dy = dx\left(a^2+x^2\right)\sqrt{a^2-x^2} + 2x^2dx\sqrt{a^2-x^2} - \frac{x^2dx\left(a^2+x^2\right)}{\sqrt{a^2-x^2}}$$

$$= \frac{dx\left(a^4+a^2x^2-4x^4\right)}{\sqrt{a^2-x^2}}.$$

164. Although the quotient of two functions can be thought of as the product of two functions, it may be more convenient to use a rule for differentiating a quotient. Let p/q be a given function whose differential we need to find. When we substitute $x + dx$ for x the quotient becomes[1]

$$\frac{p+dp}{q+dq} = (p+dp)\left(\frac{1}{q} - \frac{dq}{q^2}\right) = \frac{p}{q} - \frac{p\,dq}{q^2} + \frac{dp}{q} - \frac{dp\,dq}{q^2}.$$

When p/q is subtracted, the differential remains,

$$d.\frac{p}{q} = \frac{dp}{q} - \frac{p\,dq}{q^2},$$

since the term $dp\,dq/q^2$ vanishes. Hence, we have

$$d.\frac{p}{q} = \frac{q\,dp - p\,dq}{q^2},$$

and the rule for quotients can be stated:

> *To obtain the differential of a quotient, from the product of the denominator and the differential of the numerator we subtract the product of the numerator and the differential of the denominator. Then the remainder is divided by the square of the denominator.*

The following examples illustrate the application of this rule.

I. If $y = \dfrac{x}{a^2 - x^2}$, then by this rule

$$dy = \frac{\left(a^2+x^2\right)dx - 2x^2dx}{\left(a^2+x^2\right)^2} = \frac{\left(a^2-x^2\right)dx}{\left(a^2+x^2\right)^2}.$$

II. If $y = \dfrac{\sqrt{a^2+x^2}}{a^2-x^2}$, we have

$$dy = \frac{\left(a^2-x^2\right)x\,dx/\sqrt{a^2+x^2} + 2x\,dx\sqrt{a^2+x^2}}{(a^2-x^2)^2},$$

and when this is reduced we have

$$dy = \frac{\left(3a^2+x^2\right)x\,dx}{\left(a^2-x^2\right)^2\sqrt{a^2+x^2}}.$$

---

[1] If we wish to keep *all* terms up to the second order, the term $p\,dq^2/q^3$ cannot be omitted.

Frequently, it may be more expeditious to use the rule in its earlier form

$$d.\frac{p}{q} = \frac{dp}{q} - \frac{p\,dq}{q^2},$$

so that the differential of a quotient is equal to the quotient of the differential of the numerator by the denominator minus the quotient of the product of the differential of the denominator and the numerator by the square of the denominator. From this we have:

III. If $y = \frac{a^2 - x^2}{a^4 + a^2x^2 + x^4}$, then

$$dy = \frac{-2x\,dx}{a^4 + a^2x^2 + x^4} - \frac{\left(a^2 - x^2\right)\left(2a^2x\,dx + 4x^3dx\right)}{\left(a^4 + a^2x^2 + x^4\right)^2},$$

and when we take a common denominator we have

$$dy = \frac{-2x\,dx\left(2a^4 + 2a^2x^2 - x^4\right)}{\left(a^4 + a^2x^2 + x^4\right)^2}.$$

165. This should be sufficient for the investigation of differentials of rational functions. If the function happens to be a polynomial, we have said enough. If the function is a quotient, it can always be reduced to the following form:

$$y = \frac{A + Bx + Cx^2 + Dx^3 + Ex^4 + Fx^5 + \cdots}{\alpha + \beta x + \gamma x^2 + \delta x^3 + \epsilon x^4 + \zeta x^5 + \cdots}.$$

We let the numerator be equal to p and the denominator be equal to q, so that $y = p/q$ and

$$dy = \frac{q\,dp - p\,dq}{q^2}.$$

But since

$$p = A + Bx + Cx^2 + Dx^3 + Ex^4 + \cdots$$

and

$$q = \alpha + \beta x + \gamma x^2 + \delta x^3 + \epsilon x^4 + \cdots,$$

we have

$$dp = B\,dx + 2Cx\,dx + 3Dx^2dx + 4Ex^3dx + \cdots$$

and

$$dq = \beta\,dx + 2\gamma x\,dx + 3\delta x^2dx + 4\epsilon x^3dx + \cdots.$$

By multiplication we obtain

$$\begin{aligned} q\,dp = \alpha B\,dx + 2\alpha Cx\,dx + 3\alpha Dx^2dx &+ 4\alpha Ex^3dx + \cdots \\ +\beta Bx\,dx + 2\beta Cx^2dx &+ 3\beta Dx^3dx + \cdots \\ +\gamma Bx^2dx &+ 2\gamma Cx^3dx + \cdots \\ &+ \delta Bx^3dx + \cdots ; \end{aligned}$$

$$\begin{aligned} p\,dq = \beta A\,dx + \beta Bx\,dx + \beta Cx^2dx &+ \beta Dx^3dx + \cdots \\ +2\gamma Ax\,dx + 2\gamma Bx^2dx &+ 2\gamma Cx^3dx + \cdots \\ +3\delta Ax^2dx &+ 3\delta Bx^3dx + \cdots \\ &+ 4\epsilon Ax^3dx + \cdots . \end{aligned}$$

From these we obtain the desired differential dy, which is the quotient whose numerator is equal to

$$\begin{aligned} &(\alpha B - \beta A)\,dx + (2\alpha C - 2\gamma A)\,x\,dx + (3\alpha D + \beta C - \gamma B - 3\delta A)\,x^2dx \\ &+ (4\alpha E + 2\beta D - 2\delta B - 4\epsilon A)\,x^3dx \\ &+ (5\alpha F + 3\beta E + \gamma D - \delta C - 3\epsilon B - 5\zeta A)\,x^4dx \end{aligned}$$

and whose denominator is equal to

$$\left(\alpha + \beta x + \gamma x^2 + \delta x^3 + \epsilon x^4 + \zeta x^5 + \cdots\right)^2.$$

This expression is most accommodated to the expeditious differentiation of any rational function. Since the numerator of the differential is made up from coefficients of the numerator and denominator functions, it can be obtained by inspection. The denominator of the differential is the square of the denominator of the given function.

166. If in the given quotient either the numerator or the denominator, or both, is made up of a product, then when the multiplication is performed we have a form we have already differentiated. However, we give special rules to make it easier to cover these cases.

Suppose the given quotient has the form $y = pr/q$. We let $pr = P$. Then

$$dP = p\,dr + r\,dp.$$

Since $y = P/q$, we have

$$dy = \frac{q\,dP - P\,dq}{q^2},$$

and after substituting for P and dP, we have the following result:

I. If $y = \frac{pr}{q}$, then

$$dy = \frac{pq\,dr + qr\,dp - pr\,dq}{q^2}.$$

If $y = \frac{p}{qs}$, we let $qs = Q$, so that

$$dQ = q\,ds + s\,dq$$

and

$$dy = \frac{Q\,dp - p\,dQ}{q^2 s^2}.$$

It follows that:

II. If $y = \frac{p}{qs}$, then

$$dy = \frac{qs\,dp - pq\,ds - ps\,dq}{q^2 s^2}.$$

If $y = \frac{pr}{qs}$, again, we let $pr = P$ and $qs = Q$, so that $y = \frac{P}{Q}$ and

$$dy = \frac{Q\,dP - P\,dQ}{Q^2}.$$

Since

$$dP = p\,dr + r\,dp \qquad \text{and} \qquad dQ = q\,ds + s\,dq,$$

we obtain the following differentiation:

III. If $y = \frac{pr}{qs}$, then

$$dy = \frac{pqs\,dr + qrs\,dp - pqr\,ds - prs\,dq}{q^2 s^2},$$

or

$$dy = \frac{r\,dp}{qs} + \frac{p\,dr}{qs} - \frac{pr\,dq}{q^2 s} - \frac{pr\,ds}{qs^2}.$$

In a similar way, if the numerator and denominator of the quotient contained several factors, using the same reasoning we could investigate the differential. It does not seem to be necessary that one be led by hand through the argument. For this reason we omit any examples of this kind, especially since we will soon give a general method that will comprehend all of these special methods of differentiating.

167. There are some cases in which the differential can be more easily expressed than with the general rules we have given; these would be either products or quotients in which the factors that make up the product or the numerator or denominator of the quotient are powers.

We suppose that the function that is to be differentiated is $y = p^m q^n$. To find the differential of this function we let $p^m = P$ and $q^n = Q$, so that

$$y = PQ \qquad \text{and} \qquad dy = P\,dQ + Q\,dP.$$

Since

$$dP = mp^{m-1}dp \qquad \text{and} \qquad dQ = nq^{n-1}dq,$$

when we substitute these values, we obtain

$$dy = np^m q^{n-1}dq + mp^{m-1}q^n\,dp = p^{m-1}q^{n-1}\left(np\,dq + mq\,dp\right).$$

From this result we derive the following rule:

I. If $y = p^m q^n$, then

$$dy = p^{m-1}q^{n-1}\left(np\,dq + mq\,dp\right).$$

In a similar way, if there are three factors, the differential can be found and expressed as follows:

II. If $y = p^m q^n r^k$, then

$$dy = p^{m-1}q^{n-1}r^{k-1}\left(mqr\,dp + npr\,dq + kpq\,dr\right).$$

168. If a quotient has either a numerator or a denominator that has a factor that is a power, we can give special rules.

First we suppose that the quotient has the form $y = p^m/q$. Then from the general rule for quotients we have

$$dy = \frac{mp^{m-1}q\,dp - p^m dq}{q^2},$$

but this differential can be expressed more conveniently as:

I. If $y = \dfrac{p^m}{q}$, then

$$dy = \frac{p^{m-1}\left(mq\,dp - p\,dq\right)}{q^2}.$$

If, on the other hand, $y = p/q^n$, then by the general rule,

$$dy = \frac{q^n\,dp - npq^{n-1}dq}{q^{2n}}.$$

If we divide both numerator and denominator by q^{n-1} we have

$$dy = \frac{q\,dp - np\,dq}{q^{n+1}}.$$

We conclude:

II. If $y = \dfrac{p}{q^n}$, then

$$dy = \frac{q\,dp - np\,dq}{q^{n+1}}.$$

If the given quotient is $y = p^m/q^n$, then we find that

$$dy = \frac{mp^{m-1}q^n\,dp - np^m q^{n-1}dq}{q^{2n}},$$

which reduces to

$$dy = \frac{mp^{m-1}q\,dp - np^m\,dq}{q^{n+1}}.$$

It follows that:

III. If $y = \dfrac{p^m}{q^n}$, then

$$dy = \frac{p^{m-1}\,(mq\,dp - np\,dq)}{q^{n+1}}.$$

Finally, if the given quotient is $y = r/(p^m q^n)$, then by the general quotient rule we have

$$dy = \frac{p^m q^n\,dr - mp^{m-1}q^n r\,dp - np^m q^{n-1}r\,dq}{p^{2m}q^{2n}}.$$

Since both numerator and denominator are divisible by $p^{m-1}q^{n-1}$:

IV. If $y = \dfrac{r}{p^m q^n}$, then

$$dy = \frac{pq\,dr - mqr\,dp - npr\,dq}{p^{m+1}q^{n+1}}.$$

If several factors occur, this kind of special rule can easily be worked out, so it is superfluous to say more.

169. The rules for differentiating that we have presented so far are sufficient to cover any algebraic function of x. If the function is a sum of powers of x, this has been treated in paragraph 159; if the function is a quotient of such functions, we have shown how to differentiate in paragraph 165. We have also given an outline of differentiation when the function involves factors. We have also taught how to differentiate irrational quantities, howsoever they may affect the function, whether through addition, subtraction, multiplication, or division. We are always able to reduce the function to cases already treated. We should understand that the reference is to explicit functions. As to implicit functions given by an equation, these we will treat later, after we have taught how to differentiate functions of two or more variables.

170. If we carefully consider all of the rules we have given so far, and we compare them with each other, we can reduce them to one universal principle, which we will be able to prove rigorously in paragraph 214. In the meantime it is not so difficult to see intuitively that this is true. Any algebraic function is composed of parts that are related to each other by addition, subtraction, multiplication, or division, and these parts are either rational or irrational. We call those quantities that make up any function its parts.

We differentiate any part of a given function by itself, as if it were the only variable and the other parts were constants. Once we have the individual differentials of the parts making up the function, we put it all together in a single sum, and thus we obtain the differential of the given function.

By means of these rules almost all functions can be differentiated, not even excepting transcendental functions, as we shall show later.

171. In order to illustrate this rule, we suppose that the function consists of two parts, connected by either addition or subtraction, so that

$$y = p \pm q.$$

We suppose that the first part p is the variable part and that the second part q is the constant part, so that the differential is equal to dp. Then we suppose that the second part $\pm q$ is the only variable, while the other part p is constant, so that the differential is equal to $\pm dq$. The desired differential is put together from those two differentials, so that

$$dy = dp \pm dq,$$

just as we have seen before. From this it must be perfectly clear that if the function consists of several parts conjoined by either addition or subtraction, namely,

$$y = p \pm q \pm r \pm s,$$

then by this rule we will have

$$dy = dp \pm dq \pm dr \pm ds,$$

which is clear from the rule stated previously.

172. If the parts are joined by multiplication, so that

$$y = pq,$$

it is clear that if we suppose that only the part p is variable, then the differential will be equal to $q\,dp$. If the other part q is the only variable, then the differential is equal to $p\,dq$. When we add these two differentials we obtain the desired differential

$$dy = q\,dp + p\,dq,$$

just as we proposed above. If there are several parts joined by multiplication, for example,

$$y = pqrs,$$

and we successively let each part be variable, we obtain the differentials

$$qrs\,dp, \qquad prs\,dq, \qquad pqs\,dr, \qquad pqr\,ds,$$

whose sum gives the desired differential

$$dy = qrs\,dp + prs\,dq + pqs\,dr + pqr\,ds,$$

as we have already seen. Therefore, the differential is obtained from the differentials of all of the parts, whether they are joined by addition, subtraction, or multiplication.

173. If the parts of the function are joined by division, for example,

$$y = \frac{p}{q},$$

according to the rule we first let p be variable, and since q is constant, the differential is equal to dp/q. Next we let q alone be variable, and since $y = pq^{-1}$, the differential is equal to $-p\,dq/q^2$. When we join the two differentials we have the differential of the given function

$$dy = \frac{dp}{q} - \frac{p\,dq}{q^2} = \frac{q\,dp - p\,dq}{q^2},$$

as we have seen before. In a similar way if the given function is

$$y = \frac{pq}{rs},$$

we let each of the parts successively be variable and obtain the following differentials:

$$\frac{q\,dp}{rs}, \qquad \frac{p\,dq}{rs}, \qquad \frac{-pq\,dr}{r^2 s}, \qquad \frac{-pq\,ds}{rs^2}.$$

It follows that

$$dy = \frac{qrs\,dp + prs\,dq - pqs\,dr - pqr\,ds}{r^2 s^2}.$$

174. Provided only that each of the parts that make up a function are such that we can find their differentials, we can find the differential of the whole function. Hence, if the parts are integral powers, we can find their differentials not only by means of the laws we have given before, but also from this general rule. If the parts are irrational, since the irrationality comes from the fractional exponents, we can differentiate these through the differentiation of powers, that is, $d.x^n = nx^{n-1}dx$. From this same well we draw the differentiation of like irrational formulas that involve other surds. Therefore, it should be clear that if with this general rule, which will be proved later, we join the rule for differentiating powers, then the differentials of absolutely all algebraic functions can be exhibited.

175. From all of this it clearly follows that if y is any [algebraic] function of x, its differential dy will have the form $dy = p\,dx$, where p can always be found from the laws we have set down. Furthermore, the function p is also an algebraic function of x, since in determining the differential no other operations were used except the usual ones for algebraic functions. For this reason if y is an algebraic function of x, then dy/dx is also an algebraic function of x. Furthermore, if z is an algebraic function of x, such that $dz = q\,dx$, since q is an algebraic function of x, we also know that dz/dx is an algebraic function of x, and indeed so is dz/dy an algebraic function of x which is equal to p/q. Hence, if the formula dz/dy is part of some algebraic expression, this does not prevent the whole expression from being algebraic, provided only that y and z are algebraic functions.

176. We can extend this line of reasoning to second- and higher-order differentials. If y is an algebraic function of x, $dy = p\,dx$ and $dp = q\,dx$, then with dx remaining constant, we have $d^2y = q\,dx^2$, as we have already seen. Since for the reasons already given q is also an algebraic function of x, it follows that d^2y/dx^2 is not only a finite quantity but also an algebraic function of x, provided only that y is such a function. In a similar way we see that

$$\frac{d^3y}{dx^3}, \qquad \frac{d^4y}{dx^4}, \qquad \ldots$$

are algebraic functions of x, provided that y is such a function. Furthermore, if z is also an algebraic function of x, all finite expressions made up of differentials of any order of y, z, and dx, such as

$$\frac{d^2y}{d^2z}, \qquad \frac{d^3y}{dz\,d^2y}, \qquad \frac{dx\,d^4y}{dy^3d^2z},$$

are all likewise algebraic functions of x.

177. Since the first differential of any algebraic function of x can now be found by the method given, using the same method we can investigate the second- and higher-order differentials. If y is any algebraic function of x, from differentiation we have $dy = p\,dx$, and we note the value of p. If we differentiate again and obtain $dp = q\,dx$, then $d^2y = q\,dx^2$, supposing that dx is constant. In this way we have defined the second differential. When we differentiate q, so that $dq = r\,dx$, we have the third differential $d^3y = r\,dx^3$. In this way we investigate the differentials of higher order, and since the quantities p, q, $r, \ldots$ are all algebraic functions of x, the given laws for differentiation are sufficient. Therefore, we have continuous differentiation. If we omit the dx in the differentiation of y, we obtain the value $dy/dx = p$, which is again differentiated and divided by dx to obtain $q = d^2y/dx^2$. Each time we divide by dx, since everywhere the differential dx is omitted. In a similar way we obtain $r = d^3y/dx^3$, and so forth.

I. Let $y = \dfrac{a^2}{a^2+x^2}$; find the first- and higher-order differentials.

First we differentiate and divide by dx to obtain

$$\frac{dy}{dx} = \frac{-2a^2x}{(a^2+x^2)^2}$$

and then

$$\frac{d^2y}{dx^2} = \frac{-2a^4+6a^2x^2}{(a^2+x^2)^3},$$

$$\frac{d^3y}{dx^3} = \frac{24a^4x-24a^2x^3}{(a^2+x^2)^4},$$

$$\frac{d^4y}{dx^4} = \frac{24a^6-240a^4x^2+120a^2x^4}{(a^2+x^2)^5},$$

$$\frac{d^5y}{dx^5} = \frac{-720a^6x+2400a^4x^3-720a^2x^5}{(a^2+x^2)^6},$$

and so forth.

II. Let $y = \dfrac{1}{\sqrt{1-x^2}}$; find the first- and higher-order differentials.

$$\frac{dy}{dx} = \frac{x}{(1-x^2)^{3/2}},$$

$$\frac{d^2y}{dx^2} = \frac{1+2x^2}{(1-x^2)^{5/2}},$$

$$\frac{d^3y}{dx^3} = \frac{9x+6x^3}{(1-x^2)^{7/2}},$$

$$\frac{d^4}{dx^4} = \frac{9+72x^2+24x^4}{(1-x^2)^{9/2}},$$

$$\frac{d^5y}{dx^5} = \frac{225x+600x^3+120x^5}{(1-x^2)^{11/2}},$$

$$\frac{d^6y}{dx^6} = \frac{225+4050x^2+5400x^4+720x^6}{(1-x^2)^{13/2}},$$

and so forth. These differentials can easily be continued, but the law by which the terms proceed may not be immediately obvious. The coefficient of the highest power of x is the product of the natural numbers from 1 to the order of the differential. Meanwhile, if we wish to continue further our investigation, we will find that if $y = 1/\sqrt{1-x^2}$, generally we have

$$\begin{aligned}\frac{d^ny}{dx^n} &= \frac{1\cdot 2\cdot 3\cdots n}{(1-x^2)^{n+\frac{1}{2}}} \\ &\times \left(x^n + \frac{1}{2}\cdot\frac{n\,(n-1)}{1\cdot 2}x^{n-2}\right. \\ &\quad + \frac{1\cdot 3}{2\cdot 4}\cdot\frac{n\,(n-1)\,(n-2)\,(n-3)}{1\cdot 2\cdot 3\cdot 4}x^{n-4} \\ &\quad + \frac{1\cdot 3\cdot 5}{2\cdot 4\cdot 6}\cdot\frac{n\,(n-1)\cdots(n-5)}{1\cdot 2\cdots 6}x^{n-6} \\ &\quad \left. + \frac{1\cdot 3\cdot 5\cdot 7}{2\cdot 4\cdot 6\cdot 8}\cdot\frac{n\,(n-1)\cdots(n-7)}{1\cdot 2\cdots 8}x^{n-8} + \cdots\right).\end{aligned}$$

Examples of this kind are useful not only for acquiring a habit of differentiating, but they also provide rules that are observed in differentials of all orders, which are very much worth noticing and can lead to further discoveries.

6

On the Differentiation of Transcendental Functions

178. Besides the infinite class of transcendental, or nonalgebraic, quantities that integral calculus supplies in abundance, in *Introduction to Analysis of the Infinite* we were able to gain some knowledge of more usual quantities of this kind, namely, logarithms and circular arcs. In that work we explained the nature of these quantities so clearly that they could be used in calculation with almost the same facility as algebraic quantities. In this chapter we will investigate the differentials of these quantities in order that their character and properties can be even more clearly understood. With this understanding, a portal will be opened up into integral calculus, which is the principal source of these transcendental quantities.

179. We begin with logarithmic quantities, that is, functions of x that, besides algebraic expressions, also involve logarithms of x or any functions of logarithms of x. Since algebraic quantities no longer are a problem, the whole difficulty in finding differentials of these quantities lies in discovering the differential of any logarithm itself. There are many kinds of logarithms, which differ from each other only by a constant multiple. Here we will consider in particular the hyperbolic, or natural, logarithm, since the others can easily be found from this one. If the natural logarithm of the function p is signified by $\ln p$, then the logarithm with a different base of the same function p will be $m \ln p$ where m is a number that relates logarithms with this base to the hyperbolic logarithms. For this reason $\ln p$ will always indicate the hyperbolic logarithm of p.

180. We are investigating the differential of the hyperbolic logarithm of x and we let $y = \ln x$, so that we have to define the value of dy. We substitute $x + dx$ for x so that y is transformed into $y^{\mathrm{I}} = y + dy$. From this we have

$$y + dy = \ln(x + dx), \qquad dy = \ln(x + dx) - \ln x = \ln\left(1 + \frac{dx}{x}\right).$$

But we have seen before[1] that the hyperbolic logarithm of this kind of expression $1 + z$ can be expressed in an infinite series as follows:

$$\ln(1 + z) = z - \frac{z^2}{2} + \frac{z^3}{3} - \frac{z^4}{4} + \cdots.$$

When we substitute dx/x for z we obtain

$$dy = \frac{dx}{x} - \frac{dx^2}{2x^2} + \frac{dx^3}{3x^3} - \cdots.$$

Since all of the terms of this series vanish in the presence of the first term, we have

$$d \ln x = dy = \frac{dx}{x}.$$

It follows that the differential of any logarithm whatsoever that has the ratio to the hyperbolic logarithm of $n : 1$, has the form $n\,dx/x$.

181. Therefore, if $\ln p$ for any function p of x is given, by the same argument, we see that its differential will be dp/p. Hence, in order to find the differential of any logarithm we have the following rule:

For any quantity p whose logarithm is proposed, we take the differential of that quantity p and divide by the quantity p itself in order to obtain the desired differential of the logarithm.

This same rule follows from the form

$$\frac{p^0 - 1^0}{0},$$

to which we reduced the logarithm of p in the previous book.[2] Let $\omega = 0$, and since $\ln p = (p^\omega - 1)/\omega$, we have

$$d \ln p = d\frac{1}{\omega}p^\omega = p^{\omega - 1}dp = \frac{dp}{p},$$

since $\omega = 0$. It is to be noted, however, that dp/p is the differential of the hyperbolic logarithm of p, so that if the common logarithm of p is desired, this differential dp/p must be multiplied by the number $0.43429448\ldots$.

---

[1] *Introduction*, Book I, Chapter VII.
[2] *Introduction*, Book I, Chapter VII.

182. By means of this rule the differential of the logarithm of any given function of x whatsoever is easily found, which will be clear from the following examples.

I. If $y = \ln x$, then

$$dy = \frac{dx}{x}.$$

II. If $y = \ln x^n$, we let $x^n = p$, so that $y = \ln p$ and $dy = dp/p$. But $dp = nx^{n-1}dx$, so that

$$dy = \frac{n\,dx}{x}.$$

The same result can be found from the nature of logarithms; since $\ln x^n = n \ln x$, we have

$$d \ln x^n = nd \ln x = \frac{n\,dx}{x}.$$

III. If $y = \ln\left(1 + x^2\right)$, then

$$dy = \frac{2x\,dx}{1 + x^2}.$$

IV. If $y = \ln \dfrac{1}{\sqrt{1 - x^2}}$, since

$$y = -\ln\sqrt{1 - x^2} = -\frac{1}{2}\ln\left(1 - x^2\right),$$

we see that

$$dy = \frac{x\,dx}{1 - x^2}.$$

V. If $y = \ln \dfrac{x}{\sqrt{1 + x^2}}$, since $y = \ln x - \frac{1}{2}\ln\left(1 + x^2\right)$, we have

$$dy = \frac{dx}{x} - \frac{x\,dx}{1 + x^2} = \frac{dx}{x\left(1 + x^2\right)}.$$

VI. If $y = \ln\left(x + \sqrt{1 + x^2}\right)$, we have

$$dy = \frac{dx + x\,dx\big/\sqrt{1 + x^2}}{x + \sqrt{1 + x^2}} = \frac{x\,dx + dx\sqrt{1 + x^2}}{\left(x + \sqrt{1 + x^2}\right)\sqrt{1 + x^2}};$$

but since both numerator and denominator of this fraction are divisible by $x + \sqrt{1 + x^2}$, we have

$$dy = \frac{dx}{\sqrt{1 + x^2}}.$$

VII. If $y = \frac{1}{\sqrt{-1}} \ln\left(x\sqrt{-1} + \sqrt{1 - x^2}\right)$, we let $z = x\sqrt{-1}$. Also, since $y = \frac{1}{\sqrt{-1}} \ln\left(z + \sqrt{1 + z^2}\right)$, by the previous example we have

$$dy = \frac{1}{\sqrt{-1}} \frac{dz}{\sqrt{1 + z^2}}.$$

Since $dz = dx\sqrt{-1}$, we have

$$dy = \frac{dx}{\sqrt{1 - x^2}}.$$

Although the given logarithm involves a complex number, the differential is real.

183. If the logarithm of a product is given, then the logarithm is expressed as a sum in the following manner. If $y = \ln pqrs$ is given, since $y = \ln p + \ln q + \ln r + \ln s$, we have

$$dy = \frac{dp}{p} + \frac{dq}{q} + \frac{dr}{r} + \frac{ds}{s}.$$

This reduction also has a use if the logarithm of a quotient is to be differentiated. If

$$y = \ln \frac{pq}{rs},$$

since $y = \ln p + \ln q - \ln r - \ln s$, we have

$$dy = \frac{dp}{p} + \frac{dq}{q} - \frac{dr}{r} - \frac{ds}{s}.$$

Powers give no more difficulty. If

$$y = \ln \frac{p^m q^n}{r^\mu s^\nu},$$

since $y = m \ln p + n \ln q - \mu \ln r - \nu \ln s$, we have

$$dy = \frac{m\,dp}{p} + \frac{n\,dq}{q} - \frac{\mu\,dr}{r} - \frac{\nu\,ds}{s}.$$

I. If $y = \ln(a + x)(b + x)(c + x)$, since

$$y = \ln(a + x) + \ln(b + x) + \ln(c + x),$$

the desired differential is

$$dy = \frac{dx}{a + x} + \frac{dx}{b + x} + \frac{dx}{c + x}.$$

II. If $y = \frac{1}{2} \ln \left(\frac{1+x}{1-x} \right)$, then

$$y = \frac{1}{2} \ln (1+x) - \frac{1}{2} \ln (1-x),$$

so that

$$dy = \frac{\frac{1}{2} dx}{1+x} + \frac{\frac{1}{2} dx}{1-x} = \frac{dx}{1-x^2}.$$

III. If $y = \frac{1}{2} \ln \left(\frac{\sqrt{1+x^2}+x}{\sqrt{1+x^2}-x} \right)$, since

$$y = \frac{1}{2} \ln \left(\sqrt{1+x^2}+x \right) - \frac{1}{2} \ln \left(\sqrt{1+x^2}-x \right),$$

we have

$$dy = \frac{\frac{1}{2} dx}{\sqrt{1+x^2}} + \frac{\frac{1}{2} dx}{\sqrt{1+x^2}} = \frac{dx}{\sqrt{1+x^2}}.$$

This same result can be more easily obtained if we rationalize the denominator by multiplying both numerator and denominator by $\sqrt{1+x^2}+x$. The result is

$$y = \frac{1}{2} \ln \left(\sqrt{1+x^2}+x \right)^2 = \ln \left(\sqrt{1+x^2}+x \right),$$

and as we have seen before, $dy = dx/\sqrt{1+x^2}$.

IV. If

$$y = \ln \left(\frac{\sqrt{1+x}+\sqrt{1-x}}{\sqrt{1+x}-\sqrt{1-x}} \right),$$

we let the numerator of this fraction be

$$\sqrt{1+x}+\sqrt{1-x} = p$$

and the denominator

$$\sqrt{1+x}-\sqrt{1-x} = q,$$

so that $y = \ln \left(\frac{p}{q} \right) = \ln p - \ln q$ and $dy = dp/p - dq/q$. But

$$dp = \frac{dx}{2\sqrt{1+x}} - \frac{dx}{2\sqrt{1-x}} = \frac{-dx}{2\sqrt{1-x^2}} \left(\sqrt{1+x} - \sqrt{1-x} \right) = \frac{-q\,dx}{2\sqrt{1-x^2}}$$

and

$$dq = \frac{dx}{2\sqrt{1+x}} + \frac{dx}{2\sqrt{1-x}} = \frac{p\,dx}{2\sqrt{1-x^2}}.$$

It follows that

$$\frac{dp}{p} - \frac{dq}{q} = \frac{-q}{2p\sqrt{1-x^2}} - \frac{p\,dx}{2q\sqrt{1-x^2}} = \frac{-\left(p^2+q^2\right)dx}{2pq\sqrt{1-x^2}}.$$

Since $p^2 + q^2 = 4$ and $pq = 2x$, we have

$$dy = -\frac{dx}{x\sqrt{1-x^2}}.$$

This differential can more easily be found if the given logarithm is transformed by rationalization as follows:

$$y = \ln\frac{1+\sqrt{1-x^2}}{x} = \ln\left(\frac{1}{x} + \sqrt{\frac{1}{x^2} - 1}\right).$$

If we let

$$p = \frac{1}{x} + \sqrt{\frac{1}{x^2} - 1},$$

then

$$dp = \frac{-dx}{x^2} - \frac{dx}{x^3\sqrt{\frac{1}{x^2}-1}} = \frac{-dx}{x^2} - \frac{dx}{x^2\sqrt{1-x^2}} = \frac{-dx\left(1+\sqrt{1-x^2}\right)}{x^2\sqrt{1-x^2}}.$$

Since

$$p = \frac{1+\sqrt{1-x^2}}{x},$$

we have

$$dy = \frac{dp}{p} = \frac{-dx}{x\sqrt{1-x^2}},$$

as we have already seen.

184. Since the first differentials of logarithms, when divided by dx, are algebraic quantities, the second differentials and those of higher orders can easily be found with the rules of the previous chapter, provided that we assume that the differential dx is constant. Hence, if we let $y = \ln x$, then

$$dy = \frac{dx}{x} \quad \text{and} \quad \frac{dy}{dx} = \frac{1}{x},$$

$$d^2y = \frac{-dx^2}{x^2} \quad \text{and} \quad \frac{d^2y}{dx^2} = \frac{-1}{x^2},$$

$$d^3y = \frac{2dx^3}{x^3} \quad \text{and} \quad \frac{d^3y}{dx^3} = \frac{2}{x^3},$$

$$d^4y = \frac{-6dx^4}{x^4} \quad \text{and} \quad \frac{d^4y}{dx^4} = \frac{-6}{x^4},$$

etc. If p is an algebraic quantity and $y = \ln p$, then although y is not an algebraic quantity, nevertheless,

$$\frac{dy}{dx}, \qquad \frac{d^2y}{dx^2}, \qquad \frac{d^3y}{dx^3},$$

etc., are algebraic functions of x.

185. Now that we have discussed the differentiation of logarithms, those functions that are a combination of logarithms and algebraic functions are easily differentiated. Those functions that consist only of logarithms can also be differentiated, as is clear from the following examples.

I. If $y = (\ln x)^2$, we let $p = \ln x$, and since $y = p^2$, we have $dy = 2p\,dp$. but $dp = dx/x$, so that

$$dy = \frac{2dx}{x} \ln x.$$

II. In a similar way, if $y = (\ln x)^n$, then

$$dy = \frac{n\,dx}{x} (\ln x)^{n-1},$$

so that if $y = \sqrt{\ln x}$, since $n = \frac{1}{2}$, we have

$$dy = \frac{dx}{2x\sqrt{\ln x}}.$$

III. If p is any function of x and $y = (\ln p)^n$, then

$$dy = \frac{n\,dp}{p} (\ln p)^{n-1}.$$

Hence, since the differential dp can be found by our previous work, the differential of y itself is known.

IV. If $y = (\ln p)(\ln q)$ with p and q being any functions of x, by the product rule given before,

$$dy = \frac{dp}{p} \ln q + \frac{dq}{q} \ln p.$$

V. If $y = x \ln x$, then by the same rule,

$$dy = dx \ln x + \frac{x\,dx}{x} = dx \ln x + dx.$$

VI. If $y = x^m \ln x - \frac{1}{m}x^m$, and we differentiate each part, we see that

$$d.x^m \ln x = mx^{m-1}dx \ln x + x^{m-1}dx$$

and $d.\frac{1}{m}x^m = x^{m-1}dx$, so that

$$dy = mx^{m-1}dx \ln x.$$

VII. If $y = x^m (\ln x)^n$, then

$$dy = mx^{m-1}dx\,(\ln x)^n + nx^{m-1}dx\,(\ln x)^{n-1}.$$

VIII. If logarithms of logarithms occur, so that $y = \ln \ln x$, we let $p = \ln x$. Then $y = \ln p$ and $dy = \frac{dp}{p}$; but $dp = \frac{dx}{x}$, so that

$$dy = \frac{dx}{x \ln x}.$$

IX. If $y = \ln \ln \ln x$ and we let $p = \ln x$, then $y = \ln \ln p$, and by the preceding example

$$dy = \frac{dp}{p\,dp}.$$

But $dp = dx/x$, so that by substitution we have

$$dy = \frac{dx}{x \ln x \ln \ln x}.$$

186. Now that we have discussed the differentiation of logarithms, we move on to exponential quantities, that is, powers of the sort where the exponent is a variable. The differentials of this kind of function of x can be found by differentiating their logarithms, as follows. If we want the differential of a^x, we let $y = a^x$ and take the logarithm of each: $\ln y = x \ln a$. When we take the differentials we have $dy/y = dx \ln a$, so that $dy = y\,dx \ln a$. Since $y = a^x$, we have $dy = a^x dx \ln a$, which is the differential of a^x. In a similar way, if p is any function of x, the differential of a^p is $a^p dp \ln a$.

187. This differential could also be found immediately from the nature of exponential quantities discussed in *Introduction*. Let a^p be given where p is any function of x. When we substitute $x + dx$ for x we obtain $p + dp$. Hence, if we let $y = a^p$ and x becomes $x + dx$, we have $y + dy = a^{p+dp}$, and so

$$dy = a^{p+dp} - a^p = a^p \left(a^{dp} - 1\right).$$

We have shown[3] that an exponential quantity a^z can be expressed by a series as follows:

$$1 + z \ln a + \frac{z^2 (\ln a)^2}{2} + \frac{z^3 (\ln a)^3}{6} + \cdots .$$

It follows that

$$a^{dp} = 1 + dp \ln a + \frac{dp^2 (\ln a)^2}{2} + \cdots$$

and $a^{dp} - 1 = dp \ln a$, since the following terms vanish in the presence of $dp \ln a$. It follows that

$$dy = d.a^p = a^p dp \ln a.$$

Therefore, the differential of an exponential quantity a^p is the product of the exponential quantity itself, the differential of the exponent p, and the logarithm of the constant quantity a that is raised to the variable exponent.

188. If e is the number whose hyperbolic logarithm is equal to 1, so that $\ln e = 1$, then the differential of the quantity e^x is equal to $e^x dx$. If dx is taken to be constant, then the differential of this differential is equal to $e^x dx^2$, which is the second differential of e^x. In a similar way the third differential is equal to $e^x dx^3$. It follows that if $y = e^{nx}$, then $dy/dx = ne^{nx}$ and $d^2y/dx^2 = n^2 e^{nx}$. Furthermore,

$$\frac{d^3y}{dx^3} = n^3 e^{nx}, \qquad \frac{d^4y}{dx^4} = n^4 e^{nx}, \qquad \ldots .$$

Hence it is clear that the first, second, and following differentials of e^{nx} form a geometric progression, and it follows that the differential of order m of $y = e^{nx}$, namely, $d^m y$, equals $n^m e^{nx} dx^m$. Therefore,

$$\frac{d^m y}{y\, dx^m}$$

is the constant quantity n^m.

189. If the quantity that is raised to a power is itself a variable, its differential can be investigated in a similar way. Let p and q be any functions of x, and we consider the exponential quantity $y = p^q$. We take the logarithm so that $\ln y = q \ln p$. When we differentiate these we have

$$\frac{dy}{y} = dq \ln p + \frac{q\, dp}{p},$$

[3] *Introduction*, Book I, Chapter VII; see also note on page 1.

so that

$$dy = y\,dq \ln p + \frac{yq\,dp}{p} = p^q\,dq \ln p + qp^{q-1}dp,$$

since $y = p^q$. Hence this differential consists of two members, the first of which, $p^q dq \ln p$, would arise if in the proposed quantity p^q the p were constant and only the exponent q were variable; the other member would arise if in the proposed quantity p^q the exponent q were constant and only the quantity p were variable. This differential could have been found by the general rule given above in paragraph 170.

190. The differential of this same expression p^q can also be found from the nature of an exponential quantity as follows. Let $y = p^q$ and let x be replaced by $x + dx$ so that $y + dy = (p + dp)^{q+dq}$. This expression, when it is expressed in the usual way by a series, gives

$$y + dy = p^{q+dq} + (q + dq)\,p^{q+dq-1}dp$$
$$+ \frac{(q + dq)\,(q + dq - 1)}{1 \cdot 2} p^{q+dq-2}dp^2 + \cdots,$$

so that

$$dy = p^{q+dq} - p^q + (q + dq)\,p^{q+dq-1}dp.$$

The following terms, which involve higher powers of dp vanish in the presence of $(q + dq)\,p^{q+dq-1}dp$. But

$$p^{q+dq} - p^q = p^q \left(p^{dq} - 1\right)$$
$$= p^q \left(1 + dq \ln p + \frac{dq^2 \left(\ln p\right)^2}{2} + \cdots - 1\right) = p^q\,dq \ln p.$$

In the second term $(q + dq)\,p^{q+dq}p^{q+dq-1}dp$, if we write q instead of $q + dq$ we obtain $qp^{q-1}dp$ so that the differential is as was found before: $dy = p^q dq \ln p + qp^{q-1}dp$.

191. This same differential can more easily be investigated from the nature of exponential quantities in the following way. Since we have taken the number e for the number whose hyperbolic logarithm is equal to 1, we let $p^q = e^{q \ln p}$, because the logarithm of each is the same, $q \ln p$. Hence $y = e^{q \ln p}$. It follows, since now the quantity e that is raised to a power is constant, we have

$$dy = e^{q \ln p} \left(dq \ln p + \frac{q\,dp}{p}\right),$$

as we have shown before in the rule given in paragraph 187. When we replace $e^{q \ln p}$ with p^q, we have

$$dy = p^q \, dq \ln p + \frac{p^q q \, dp}{p} = p^q \, dq \ln p + q p^{q-1} dp.$$

If $y = x^x$, we have $dy = x^x dx \ln x$. From this its higher differentials can be defined. We see that

$$\frac{d^2y}{dx^2} = x^x \left(\frac{1}{x} + (1 + \ln x)^2 \right),$$

$$\frac{d^3y}{dx^3} = x^x \left((1 + \ln x)^3 + \frac{3\,(1 + \ln x)}{x} - \frac{1}{x^2} \right),$$

etc.

192. Among the differentials of this kind of function, which involve exponential functions, the following examples should be especially noted. They arise from the differentiation of $e^x p$, indeed,

$$d.e^x p = e^x \, dp + e^x p \, dx = e^x \,(dp + p \, dx)\,.$$

I. If $y = e^x x^n$, then

$$dy = e^x n x^{n-1} dx + e^x x^n \, dx,$$

or,

$$dy = e^x \, dx \left(n x^{n-1} + x^n \right).$$

II. If $y = e^x \,(x - 1)$, then

$$dy = e^x x \, dx.$$

III. If $y = e^x \left(x^2 - 2x + 2 \right)$, then

$$dy = e^x x^2 dx.$$

IV. If $y = e^x \left(x^3 - 3x^2 + 6x - 6 \right)$, then

$$dy = e^x x^3 dx.$$

193. If the exponents themselves are again exponential quantities, then differentiation is accomplished according to the same rules. Thus, if we want to differentiate e^{e^x}, we let $p = e^x$, so that

$$y = e^{e^x} = e^p;$$

Then $dy = e^p dp$, and since $dp = e^x dx$, it follows that if $y = e^{e^x}$, then

$$dy = e^{e^x} e^x dx.$$

If $y = e^{e^{e^x}}$, then

$$dy = e^{e^{e^x}} e^{e^x} e^x dx.$$

However, if $y = p^{q^r}$, then we let $q^r = z$, and $dy = p^z dz \ln p + zp^{z-1} dp$, but $dz = q^r dr \ln q + rq^{r-1} dq$, so that

$$dy = p^z q^r dr \ln p \cdot \ln q + p^z r q^{r-1} dq \ln p + \frac{p^z q^r dp}{p}.$$

It follows that if $y = p^{q^r}$, then

$$dy = p^{q^r} q^r \left(dr \ln p \cdot \ln q + \frac{r\, dq \ln p}{q} + \frac{dp}{p} \right).$$

In this way, no matter how the exponential may occur, the differential can be found.

194. We proceed now to transcendental quantities. Previously, a consideration of circular arcs has led us to a knowledge of these. Let an arc of a circle whose radius is always equal to unity be given, and let the sine of this arc be equal to x. We express this arc as $\arcsin x$ and we investigate the differential of this arc, that is, the increment that it receives if the sine of x is increased by its differential dx. We can accomplish this by the differentiation of logarithms, since in *Introduction* (*loc. cit.*, paragraph 138) we have shown that the expression $\arcsin x$ can be reduced to this logarithmic expression:

$$\frac{1}{\sqrt{-1}} \ln \left(\sqrt{1 - x^2} + x\sqrt{-1} \right).$$

We let $y = \arcsin x$, so that

$$y = \frac{1}{\sqrt{-1}} \ln \left(\sqrt{1 - x^2} + x\sqrt{-1} \right),$$

whose differential we have seen (paragraph 182, VII) to be

$$dy = \frac{\frac{1}{\sqrt{-1}} \left(\frac{-x\,dx}{\sqrt{1-x^2}} + dx\sqrt{-1} \right)}{\sqrt{1 - x^2} + x\sqrt{-1}} = \frac{dx \left(x\sqrt{-1} + \sqrt{1 - x^2} \right)}{\left(\sqrt{1 - x^2} + x\sqrt{-1} \right) \sqrt{1 - x^2}},$$

so that

$$dy = \frac{dx}{\sqrt{1 - x^2}}.$$

195. This differential of a circular arc can also more easily be found without the aid of logarithms. If $y = \arcsin x$, then x is the sine of the arc y, that is, $x = \sin y$. When we substitute $x + dx$ for x, y becomes $y + dy$, so that $x + dx = \sin(y + dy)$. Since

$$\sin(a + b) = \sin a \cdot \cos b + \cos a \cdot \sin b,$$

we have

$$\sin(y + dy) = \sin y \cdot \cos dy + \cos y \cdot \sin dy.$$

As dy vanishes the arc becomes equal to its sine, and its cosine becomes equal to 1. For this reason $\sin(y + dy) = \sin y + dy \cos y$, so that $x + dx = \sin y + dy \cos y$. Since $\sin y = x$, we have $\cos y = \sqrt{1 - x^2}$, and when these values are substituted, we have $dx = dy\sqrt{1 - x^2}$, from which we obtain

$$dy = \frac{dx}{\sqrt{1 - x^2}}.$$

The arc of a given sine has a differential equal to the differential of the sine divided by the cosine.

196. Suppose p is any function of x and that y is the arc whose sine is p, that is, $y = \arcsin p$. Since the differential of this arc is $dy = dp/\sqrt{1 - p^2}$, where $\sqrt{1 - p^2}$ expresses the cosine of that same arc, we can find the differential of an arc whose cosine is given. If $y = \arccos x$, then the sine of this arc is equal to $\sqrt{1 - x^2}$, so that $y = \arcsin \sqrt{1 - x^2}$. When we let $p = \sqrt{1 - x^2}$, it follows that $dp = -dx/\sqrt{1 - x^2}$ and $\sqrt{1 - p^2} = x$, so that

$$dy = \frac{-dx}{\sqrt{1 - x^2}}.$$

The differential of the arc of a given cosine is equal to the negative of the differential of the cosine divided by the sine of that same arc.

This result can also be shown in the following way. If $y = \arccos x$, we let $z = \arcsin x$, so that $dz = dx/\sqrt{1 - x^2}$. But the sum of the two arcs y and z is equal to the constant 90 degrees, that is $y + z$ is constant, so that $dy + dz = 0$, or $dy = -dz$. Hence we have the same result as before, $dy = -dx/\sqrt{1 - x^2}$.

197. If an arc whose tangent is given is to be differentiated, we begin with $y = \arctan x$. But then the sine is equal to $x/\sqrt{1 + x^2}$ and the cosine is equal to $1/\sqrt{1 + x^2}$. We let $p = x/\sqrt{1 + x^2}$, so that

$$\sqrt{1 - p^2} = \frac{1}{\sqrt{1 + x^2}},$$

and so $y = \arcsin p$. Hence, by a rule already given $dy = dp/\sqrt{1 - p^2}$. Since $p = x/\sqrt{1 + x^2}$, we have

$$dp = \frac{dx}{(1 + x^2)^{3/2}},$$

and after substitution we obtain

$$dy = \frac{dx}{1+x^2}.$$

The differential of the arc whose tangent is given is equal to the differential of the tangent divided by the square of the secant. We note that $\sqrt{1+x^2}$ is the secant if x is the tangent.

198. In a similar way, suppose that the cotangent of an arc is given, so that y is equal to the arccotangent of x. Since the tangent of that same arc is $1/x$, we let $p = 1/x$, so that $y = \tan p$. Since

$$dy = \frac{dp}{1+p^2} \qquad \text{and} \qquad dp = \frac{-dx}{x^2},$$

we make the substitutions to obtain

$$dy = \frac{-dx}{1+x^2},$$

that is, *the differential of the arc of a cotangent is the negative of the differential of the cotangent divided by the square of the cosecant.*

If y is equal to the arcsecant of x, since

$$y = \arccos \frac{1}{x},$$

we have

$$dy = \frac{dx}{x^2\sqrt{1-1/x^2}} = \frac{dx}{x\sqrt{x^2-1}}.$$

Also, if y is equal to the arccosecant of x, then

$$y = \arcsin \frac{1}{x}$$

and

$$dy = \frac{-dx}{x\sqrt{x^2-1}}.$$

Frequently, the versed sine[4] occurs. If y is equal to the versed sine of x, since $y = \cos(1-x)$, the sine of this arc is equal to $\sqrt{2x-x^2}$, so that

$$dy = \frac{dx}{\sqrt{2x-x^2}}.$$

[4]The versed sine of α is equal to $1 - \cos\alpha$.

199. Although the arc whose sine or cosine or tangent or cotangent or secant or cosecant or finally versed sine is given is a transcendental quantity, nevertheless its differential when divided by dx is an algebraic quantity. It follows that its second differential, its third, fourth, and so forth, when divided by the appropriate power of dx, are also algebraic. In order that this differentiation might better be seen, we adjoin some examples.

I. If $y = \arcsin 2x\sqrt{1-x^2}$, we let $p = 2x\sqrt{1-x^2}$, so that $y = \arcsin p$ and $dy = dp/\sqrt{1-p^2}$. But then

$$dp = 2dx\sqrt{1-x^2} - \frac{2x^2dx}{\sqrt{1-x^2}} = \frac{2dx\left(1-2x^2\right)}{\sqrt{1-x^2}}$$

and $\sqrt{1-p^2} = 1-2x^2$, so when these values are substituted we have

$$dy = \frac{2dx}{\sqrt{1-x^2}}.$$

From this it is clear that $2x\sqrt{1-x^2}$ is the sine of twice the arc where x is the sine of the original arc. Hence if $y = 2\arcsin x$, then $dy = 2dx/\sqrt{1-x^2}$.

II. If

$$y = \arcsin \frac{1-x^2}{1+x^2},$$

we let

$$p = \frac{1-x^2}{1+x^2},$$

so that

$$dp = \frac{-4x\,dx}{\left(1+x^2\right)^2} \qquad \text{and} \qquad \sqrt{1-p^2} = \frac{2x}{1+x^2}.$$

Since

$$dy = \frac{dp}{\sqrt{1-p^2}},$$

we have

$$dy = \frac{-2dx}{1+x^2}.$$

III. If $y = \arcsin\sqrt{(1-x)/2}$, we let $p = \sqrt{(1-x)/2}$, so that

$$\sqrt{1-p^2} = \sqrt{\frac{1+x}{2}} \qquad \text{and} \qquad dp = \frac{-dx}{4\sqrt{\frac{1-x}{2}}}.$$

It follows that

$$dy = \frac{dp}{\sqrt{1-p^2}} = \frac{-dx}{2\sqrt{1-x^2}}.$$

IV. If

$$y = \arctan \frac{2x}{1 - x^2},$$

we let

$$p = \frac{2x}{1 - x^2},$$

so that

$$1 + p^2 = \frac{\left(1 + x^2\right)^2}{\left(1 - x^2\right)^2} \qquad \text{and} \qquad dp = \frac{2dx\left(1 + x^2\right)}{\left(1 - x^2\right)^2}.$$

Hence, since $dy = dp/\left(1 + p^2\right)$, we have by the rule for tangents (paragraph 197)

$$dy = \frac{2dx}{1 + x^2}.$$

V. If

$$y = \arctan \frac{\sqrt{1 + x^2} - 1}{x},$$

we let

$$p = \frac{\sqrt{1 + x^2} - 1}{x},$$

so that

$$p^2 = \frac{2 + x^2 - 2\sqrt{1 + x^2}}{x^2}$$

and

$$1 + p^2 = \frac{2 + 2x^2 - 2\sqrt{1 + x^2}}{x^2} = \frac{2\left(\sqrt{1 + x^2} - 1\right)\sqrt{1 + x^2}}{x^2}.$$

But

$$dp = \frac{-dx}{x^2\sqrt{1 + x^2}} x^2 + \frac{dx}{x^2} = \frac{dx\left(\sqrt{1 + x^2} - 1\right)}{x^2\sqrt{1 + x^2}}.$$

Hence, since $dy = dp/\left(1 + p^2\right)$, we have

$$dy = \frac{dx}{2\left(1 + x^2\right)}.$$

From this we see that

$$\arctan \frac{\sqrt{1 + x^2} - 1}{x} = \frac{1}{2} \arctan x.$$

VI. If $y = e^{\arcsin x}$, this formula can also be differentiated by the preceding methods. Indeed, we have

$$dy = e^{\arcsin x} \frac{dx}{\sqrt{1 - x^2}}.$$

In this way all functions of x involving not only logarithms and exponentials, but also even circular arcs, can be differentiated.

200. Since the differentials of arcs when divided by dx are algebraic quantities, their second, and higher, differentials can be found, as we have shown, by differentiation of algebraic quantities. Let $y = \arcsin x$. Since $dy = dx/\sqrt{1 - x^2}$, we have

$$\frac{dy}{dx} = \frac{1}{\sqrt{1 - x^2}},$$

whose differential gives the value of d^2y/dx^2, provided that we keep dx constant. Hence the differentials of this y of any order are of this kind.

If $y = \arcsin x$, then

$$\frac{dy}{dx} = \frac{1}{\sqrt{1 - x^2}},$$

and when we keep dx constant,

$$\frac{d^2y}{dx^2} = \frac{x}{(1 - x^2)^{3/2}},$$

$$\frac{d^3y}{dx^3} = \frac{1 + 2x^2}{(1 - x^2)^{5/2}},$$

$$\frac{d^4y}{dx^4} = \frac{9x + 6x^3}{(1 - x^2)^{7/2}},$$

$$\frac{d^5y}{dx^5} = \frac{9 + 72x^2 + 24x^4}{(1 + x^2)^{9/2}},$$

$$\frac{d^6y}{dx^6} = \frac{225x + 600x^3 + 120x^5}{(1 - x^2)^{11/2}},$$

$$\cdots .$$

Hence we conclude as above (paragraph 177) that the general formula will be

$$\frac{d^{n+1}y}{dx^{n+1}} = \frac{1 \cdot 2 \cdot 3 \cdots n}{(1-x^2)^{n+1/2}}$$
$$\times \left(x^n + \frac{1}{2} \cdot \frac{n(n-1)}{1 \cdot 2} x^{n-2} \right.$$
$$+ \frac{1 \cdot 3}{2 \cdot 4} \cdot \frac{n(n-1)(n-2)(n-3)}{1 \cdot 2 \cdot 3 \cdot 4} x^{n-4}$$
$$+ \frac{1 \cdot 3 \cdot 5}{2 \cdot 4 \cdot 6} \cdot \frac{n(n-1)(n-2)(n-3)(n-4)(n-5)}{1 \cdot 2 \cdot 3 \cdot 4 \cdot 5 \cdot 6} x^{n-6}$$
$$\left. + \cdots \right).$$

201. There remain some quantities that arise as inverses of these functions, namely the sines and tangents of given arcs, and we ought to show how these are differentiated. Let x be a circular arc and let $\sin x$ denote its sine, whose differential we are to investigate. We let $y = \sin x$ and replace x by $x + dx$ so that y becomes $y + dy$. Then $y + dy = \sin(x + dx)$ and

$$dy = \sin(x + dx) - \sin x.$$

But

$$\sin(x + dx) = \sin x \cdot \cos dx + \cos x \cdot \sin dx,$$

and since, as we have shown in *Introduction*,

$$\sin z = \frac{z}{1} - \frac{z^3}{1 \cdot 2 \cdot 3} + \frac{z^5}{1 \cdot 2 \cdot 3 \cdot 4 \cdot 5} - \cdots,$$

$$\cos z = 1 - \frac{z^2}{1 \cdot 2} + \frac{z^4}{1 \cdot 2 \cdot 3 \cdot 4} - \cdots,$$

when we exclude the vanishing terms, we have $\cos dx = 1$ and $\sin dx = dx$, so that

$$\sin(x + dx) = \sin x + dx \cos x.$$

Hence, when we let $y = \sin x$, we have

$$dy = dx \cos x.$$

Therefore, the differential of the sine of any arc is equal to the product of the differential of the arc and the cosine of the arc.

If p is any function of x, then in a similar way we have

$$d.\sin p = dp\cos p.$$

202. Similarly, if we are given $\cos x$, that is, the cosine of the arc x, and we are to investigate its differential, we let $y = \cos x$ and replace x by $x + dx$ so that $y + dy = \cos(x + dx)$. Since

$$\cos(x + dx) = \cos x \cdot \cos dx - \sin x \cdot \sin dx,$$

and since, as we have just seen, $\cos dx = 1$ and $\sin dx = dx$, we have

$$y + dy = \cos x - dx\sin x,$$

so that

$$dy = -dx\sin x.$$

Hence, the differential of the cosine of any arc is equal to the negative of the product of the differential of the arc and the sine of the same arc.

Hence, if p is any function of x, then

$$d.\cos p = -dp\sin p.$$

These differentiations can also be derived from previous results as follows. If $y = \sin p$, then $p = \arcsin y$ and

$$dp = \frac{dy}{\sqrt{1 - y^2}}.$$

Since $y = \sin p$, $\cos p = \sqrt{1 - y^2}$, and we substitute this value to obtain $dp = dy/\cos p$, and so

$$dy = dp\cos p$$

as before. In like manner, if $y = \cos p$, then $\sqrt{1 - y^2} = \sin p$ and $p = \arccos y$. Hence

$$dp = \frac{-dy}{\sqrt{1 - y^2}} = \frac{-dy}{\sin p},$$

so that we have as before

$$dy = -dp\sin p.$$

203. If $y = \tan x$, then

$$dy = \tan(x + dx) - \tan x;$$

since

$$\tan(x+dx) = \frac{\tan x + \tan dx}{1 - \tan x \cdot \tan dx},$$

when the tangent is subtracted from this expression there remains

$$dy = \frac{\tan dx\,(1 + \tan x \cdot \tan x)}{1 - \tan x \cdot \tan dx}.$$

However, when the arc dx vanishes, the tangent is equal to the arc itself, so that $\tan dx = dx$, and the denominator $1 - dx \tan x$ reduces to unity. Hence

$$dy = dx\left(1 + \tan^2 x\right).$$

Since

$$1 + \tan^2 x = \sec^2 x = \frac{1}{\cos^2 x},$$

we have

$$dy = dx \sec^2 x = \frac{dx}{\cos^2 x}.$$

We could also obtain this differential from the differentials of the sine and cosine. Since $\tan x = \sin x/\cos x$, we have (paragraph 164)

$$dy = \frac{dx \cos x \cdot \cos x + dx \sin x \cdot \sin x}{\cos^2 x} = \frac{dx}{\cos^2 x}$$

since $\sin^2 x + \cos^2 x = 1$.

204. This differential can also be found in a different way. Since $y = \tan x$, we have $x = \arctan y$, and by the rule given above,

$$dx = \frac{dy}{1+y^2}.$$

Since $y = \tan x$,

$$\sqrt{1+y^2} = \sec x = \frac{1}{\cos x},$$

so that $dx = dy \cos^2 x$ and

$$dy = \frac{dx}{\cos^2 x}$$

as before. *The differential of the tangent of any arc is equal to the differential of the arc divided by the square of the cosine of the same arc.*

In a similar way if we let $y = \cot x$, then x is equal to the arccotangent of y and

$$dx = \frac{-dy}{1+y^2}.$$

But

$$\sqrt{1+y^2} = \csc x = \frac{1}{\sin x},$$

so that $dx = -dy \sin^2 x$ and

$$dy = \frac{-dx}{\sin^2 x}.$$

The differential of the cotangent of any arc is equal to the negative of the differential of the arc divided by the square of the sine of the same arc.

Or since

$$\cot x = \frac{\cos x}{\sin x},$$

we have from the quotient rule

$$dy = \frac{-dx \sin^2 x - dx \cos^2 x}{\sin^2 x} = \frac{-dx}{\sin^2 x}$$

as we have already seen.

205. If the secant of an arc is given, so that $y = \sec x$, since $y = 1/\cos x$, we have

$$dy = \frac{dx \sin x}{cos^2 x} = dx \tan x \sec x.$$

In a similar way, if $y = \csc x$, we have

$$dy = \frac{-dx \cos x}{sin^2 x} = -dx \cot x \csc x,$$

and it is not necessary to consider rules for special cases. If the versed sine is given and y is equal to the versed sine of x, since $y = 1 - \cos x$, we have $dy = dx \sin x$. In all of the cases in which some straight line is related to a given arc, since it can always be expressed through a sine or a cosine, it can always be differentiated without difficulty. This is true not only of the first differentials, but also of the second and succeeding differentials by the given rules. We let $y = \sin x$, $z = \cos x$, and we keep dx constant. Then we have as follows:

$$\begin{aligned} y &= \sin x, & z &= \cos x, \\ dy &= dx \cos x, & dz &= -dx \sin x, \\ d^2 y &= -dx^2 \sin x, & d^2 z &= -dx^2 \cos x, \\ d^3 y &= -dx^3 \cos x, & d^3 z &= dx^3 \sin x, \\ d^4 y &= -dx^4 \sin x, & d^4 z &= dx^4 \cos x, \\ &\dots . \end{aligned}$$

206. In a similar way we can find the differentials of all orders of the tangent of the arc x. Let $y = \tan x = \sin x / \cos x$ and keep dx constant. Then

$$\begin{aligned} y &= \frac{\sin x}{\cos x}, \\ \frac{dy}{dx} &= \frac{1}{\cos^2 x}, \\ \frac{d^2 y}{dx^2} &= \frac{2 \sin x}{\cos^3 x}, \\ \frac{d^3 y}{dx^3} &= \frac{6}{\cos^4 x} - \frac{4}{\cos^2 x}, \\ \frac{d^4 y}{dx^4} &= \frac{24 \sin x}{\cos^5 x} - \frac{8 \sin x}{\cos^3 x}, \\ \frac{d^5 y}{dx^5} &= \frac{120}{\cos^6 x} - \frac{120}{\cos^4 x} + \frac{16}{\cos^2 x}, \\ \frac{d^6 y}{dx^6} &= \frac{720 \sin x}{\cos^7 x} - \frac{480 \sin x}{\cos^5 x} + \frac{32 \sin x}{\cos^3 x}, \\ \frac{d^7 y}{dx^7} &= \frac{5040}{\cos^8 x} - \frac{6720}{\cos^6 x} + \frac{2016}{\cos^4 x} - \frac{64}{\cos^2 x}. \end{aligned}$$

207. Any function whatsoever in which the sine or cosine of an arc is involved can be differentiated by these rules. This can be seen from the following examples.

I. If $y = 2 \sin x \cos x = \sin 2x$, then

$$dy = 2dx \cos^2 x - 2dx \sin^2 x = 2dx \cos 2x.$$

II. If $y = \sqrt{\dfrac{1 - \cos x}{2}}$, or $y = \sin \frac{1}{2} x$, then

$$dy = \frac{dx \sin x}{2\sqrt{2(1 - \cos x)}}.$$

Since $\sqrt{2}\,(1-\cos x) = 2\sin\frac{1}{2}x$ and $\sin x = 2\sin\frac{1}{2}x\cos\frac{1}{2}x$, we have

$$dy = \frac{1}{2}dx\cos\frac{1}{2}x,$$

which follows immediately from the form $y = \sin\frac{1}{2}x$.

III. If $y = \cos\ln\frac{1}{x}$, we let $p = \ln\frac{1}{x}$ so that $y = \cos p$. Hence

$$dy = -dp\sin p.$$

But since $p = \ln 1 - \ln x$, we have $dp = -dx/x$, and so

$$dy = \frac{dx}{x}\sin\ln\frac{1}{x}.$$

IV. If $y = e^{\sin x}$, we have

$$dy = e^{\sin x}dx\cos x.$$

V. If $y = e^{-n/\cos x}$, then

$$dy = -\frac{e^{-n/\cos x}n\,dx\sin x}{\cos^2 x}.$$

VI. If $y = \ln\left(1-\sqrt{1-e^{-n/\sin x}}\right)$, we let $p = e^{-n/\sin x}$, and since

$$y = \ln\left(1-\sqrt{1-p}\right),$$

we have

$$dy = \frac{dp}{2\left(1-\sqrt{1-p}\right)\sqrt{1-p}}.$$

But

$$dp = \frac{e^{-n/\sin x}n\,dx\cos x}{\sin^2 x}.$$

When this value is substituted, we obtain

$$dy = \frac{ne^{-n/\sin x}dx\cos x}{2\sin^2 x\left(1-\sqrt{1-e^{-n/\sin x}}\right)\sqrt{1-e^{-n/\sin x}}}.$$

7

On the Differentiation of Functions of Two or More Variables

208. If two or more variable quantities x, y, z are independent of each other, it can happen that while one of the variables increases or decreases, the other variables remain constant. Since we have supposed that there is no connection between the variables, a change in one does not affect the others. Neither do the differentials of the quantities y and z depend on the differential of x, with the result that when x is increased by its differential dx, the quantities y and z can either remain the same, or they can change in any desired way. Hence, if the differential of x is dx, the differentials of the remaining quantities, dy and dz, remain indeterminate and by our arbitrary choice will be presumed to be either practically nothing or infinitely small when compared to dx.

209. However, frequently the letters y and z are wont to signify functions of x that are either unknown or whose relationship to x is not considered. In this case the differentials dy and dz do have a certain relationship to dx. Whether or not y and z depend on x, the method of differentiation that we now consider is the same. We look for the differential of a function that is formed in any way from the several variables x, y, and z that the function receives when each variable x, y, z increases by its respective differential dx, dy, or dz. In order to find this for the given function, for each of the variables x, y, and z we write $x + dx$, $y + dy$, and $z + dz$, and from this expression we subtract the given function. The remainder is the desired differential. This should be clear from the nature of differentials.

210. Let X be a function of x, and let its differential, or increase, be equal to $P\,dx$, when x increases by dx. Then let Y be a function of y and let its differential be equal to $Q\,dy$, which is the augmentation Y receives when y is increased to $y + dy$. Finally, let Z be a function or z, and let its differential be equal to $R\,dz$. These differentials $P\,dx$, $Q\,dy$, and $R\,dz$ can be found from the nature of the functions X, Y, and Z by means of the rules we have given above. Suppose the given function is $X+Y+Z$, which is a function of the three variables x, y, and z, and its differential is equal to $P\,dx + Q\,dy + R\,dz$. Whether these three differentials are homogeneous or not need not concern us. Terms that contain powers of dx will vanish in the presence of $P\,dx$, as if the other members $Q\,dy$ and $R\,dz$ were absent. For a similar reason we neglect terms in the differentiation of the functions Y and Z.

211. We keep the same description of X, Y, and Z and let the given function be XYZ, which is a function of x, y, and z. We investigate the differential of this function. If we replace x by $x+dx$, y by $y+dy$, and z by $z + dz$, then X becomes $X + P\,dx$, Y becomes $Y + Q\,dy$, and Z becomes $Z + R\,dz$, so that the given function XYZ becomes

$$(X + P\,dx)\,(Y + Q\,dy)\,(Z + R\,dz)$$
$$= XYZ + YZP\,dx + XZQ\,dy + XYR\,dz$$
$$+ ZPQ\,dx\,dy + YPR\,dx\,dz + XQR\,dy\,dz + PQR\,dx\,dy\,dz.$$

Since dx, dy, and dz are infinitely small, whether they are mutually homogeneous or not, the last term will vanish in the presence of any one of the preceding terms. Then the term $ZPQ\,dx\,dy$ will vanish in the presence of either $YZP\,dx$ or $XZQ\,dy$. For the same reason $YPR\,dx\,dz$ and $XQR\,dy\,dz$ will vanish. When we subtract the given function, the remainder is the differential

$$YZP\,dx + XZQ\,dy + XYR\,dz.$$

212. These examples of functions of three variables x, y, and z, to which we could, if desired, add more, should be sufficient to show that for any proposed function of three variables x, y, and z, howsoever these variables may be combined, the differential will always have this same form $p\,dx + q\,dy + r\,dz$. These functions p, q and r will be individual functions of either all three variables x, y, and z or of two variables, or even of only one, depending of how the given function is formed from the three variables and constants. In a similar way, if the given function depends on four or more variables, say x, y, z, and v, then its differential will have the form

$$p\,dx + q\,dy + r\,dz + s\,dv.$$

213. Let us consider first a function of only two variables x and y, which we will call V. Then its differential will have the form

$$dV = p\,dx + q\,dy.$$

If we let the quantity y remain constant, then $dy = 0$, so that the differential of the function V will be $p\,dx$. If, on the other hand, we let x remain constant, then $dx = 0$ and only y remains variable, so that the differential of V is equal to $q\,dy$. The result is that the rule for differentiating a function V of two variables x and y is as follows:

Let only the first quantity x remain variable, while the second quantity y is treated as a constant. We take the differential of V, which will be equal to $p\,dx$. Then we let only the quantity y remain variable, while the other, x, is kept as constant and the differential is found, which will be equal to $q\,dy$. From these results, when we let both x and y be variable we have $dV = p\,dx + q\,dy$.

214. In a similar way, when the function V is of three variables x, y, and z, the differential of this function has the form

$$dV = p\,dx + q\,dy + r\,dz.$$

It is clear that if only the quantity x is kept variable and the remaining y and z are kept constant, since $dy = 0$ and $dz = 0$, the differential of V will be equal to $p\,dx$. In a like manner we find the differential of V to be equal to $q\,dy$ when x and z are constant while only y is variable. If x and y are treated as constants and only z is variable, we see that the differential of V is equal to $r\,dz$. Hence, in order to find the differential of a function of three or more variables, we consider individually each variable as if it alone were variable and then take the differential, considering the other quantities as constant. Then we take the sum of each of these differentials found with each individual quantity taken as variable. This sum will be the required differential of the given function.

215. In this rule, which we have found for the differentiation of a function of however many variables, we have a demonstration of the general rule given above (paragraph 170) by means of which any function of one variable can be differentiated. If for each term of the function discussed in that place the variable is considered to be a different letter and then each term is successively differentiated in the way we have just prescribed, as if only that were variable, we then collect into one sum each of the differentials obtained in this way. This sum will be the desired differential after each of the individual letters has its value restored. Hence this rule has wide application to functions of several variables, no matter how they may be composed. Thus its use in all of differential calculus is quite wide.

216. Now that we have found this general rule, by means of which functions of howsoever many variables can be differentiated, it will be pleasant to show its use in several examples.

I. If $V = xy$, then

$$dV = x\,dy + y\,dx.$$

II. If $V = \frac{x}{y}$, then

$$dV = \frac{dx}{y} - \frac{x\,dy}{y^2}.$$

III. If $V = \frac{y}{\sqrt{a^2 - x^2}}$, then

$$dV = \frac{dy}{\sqrt{a^2 - x^2}} + \frac{yx\,dx}{\left(a^2 - x^2\right)^{3/2}}.$$

IV. If $V = (\alpha x + \beta y + \gamma)^m (\delta x + \epsilon y + \zeta)^n$, then

$$\begin{aligned} dV &= m(\alpha x + \beta y + \gamma)^{m-1} (\delta x + \epsilon y + \zeta)^n (\alpha dx + \beta dy) \\ &\quad + n(\alpha x + \beta y + \gamma)^m (\delta x + \epsilon y + \zeta)^{n-1} (\delta dx + \epsilon dy), \end{aligned}$$

or

$$dV = (\alpha x + \beta y + \gamma)^{m-1} (\delta + \epsilon y + \zeta)^{n-1}$$

by

$$\begin{aligned} &(m\alpha\delta + n\alpha\delta)\,x\,dx + (m\beta\delta + n\alpha\epsilon)\,x\,dy + (m\alpha\epsilon + n\beta\delta)\,y\,dx \\ &+ (m\beta\epsilon + n\beta\epsilon)\,y\,dy + (m\alpha\zeta + n\gamma\delta)\,dx + (m\beta\zeta + n\gamma\epsilon)\,dy. \end{aligned}$$

V. If $V = y \ln x$, then

$$dV = dy \ln x + \frac{y\,dx}{x}.$$

VI. If $V = x^y$, then

$$dV = yx^{y-1}dx + x^y\,dy \ln x.$$

VII. If $V = \arctan \frac{y}{x}$, then

$$dV = \frac{x\,dy - y\,dx}{x^2 + y^2}.$$

VIII. If $V = \sin x \cos y$, then

$$dV = dx \cos x \cos y - dy \sin x \sin y.$$

IX. If $V = \dfrac{e^x y}{\sqrt{x^2+y^2}}$, then

$$dV = \frac{e^z y\, dz}{\sqrt{x^2+y^2}} + \frac{e^z\left(x^2 dy - yx\, dx\right)}{\left(x^2+y^2\right)\sqrt{x^2+y^2}}.$$

X. If $V = e^z \arcsin\left(\dfrac{x-\sqrt{x^2-y^2}}{x+\sqrt{x^2-y^2}}\right)$, the result is

$$dV = e^z dz \arcsin\left(\frac{x-\sqrt{x^2-y^2}}{x+\sqrt{x^2-y^2}}\right)$$

$$+ e^z \frac{xy\, dy - y^2 dx}{\left(x+\sqrt{x^2-y^2}\right)\left(x^2-y^2\right)^{3/4}\sqrt{x}}.$$

217. We have seen that if V is a function of two variables x and y, its differential will have the form $dV = P\, dx + Q\, dy$, in which P and Q are functions that depend on V and are determined by it. It follows that these two quantities P and Q in some certain way depend on each other, since each depends on the function V. Whatever this connection between the finite quantities P and Q may be, which we will have to investigate, it is clear that not all differential formulas with the form $P\, dx + Q\, dy$, in which P and Q are arbitrarily chosen, can be the differential of some finite function V of x and y. Unless this relationship between the functions P and Q is present, which the nature of differentiation requires, a differential of the type $P\, dx + Q\, dy$ clearly cannot arise from differentiation, and in turn will have no integral.

218. Therefore, in integration it is of great interest to know this relationship between the quantities P and Q in order that we may distinguish between those that really arose from the differentiation of some finite function and those that were formed arbitrarily and have no integral. Although we are not going to take up the business of integration here, still this is a convenient time to investigate this relationship by looking more deeply into the nature of real differentials. Not only is this knowledge extremely necessary for integral calculus, for which we are preparing the way, but also it will cast significant light on differential calculus itself. First, then, it is clear that if V is a function of two variables x and y, then both the differentials dx and dy must be present in the differential $P\, dx + Q\, dy$. It follows that it is not possible for P to be equal to zero, nor for Q to be equal to zero. Hence, if P is a function of x and y, the formula $P\,dx$ cannot be the differential of some finite quantity, that is, there is no finite quantity whose differential is $P\,dx$.

219. Thus there is no finite quantity V, whether algebraic or transcendental, whose differential is $yx\,dx$, provided that y is a variable quantity which is independent of x. If we should suppose that there did exist such a finite quantity V, since y is part of its differential, it would necessarily be the case that y would also be in the quantity V. But if V contained y, due to the variability of y, necessarily dy would have to be in the differential of V. However, since it is not present, it is not possible that $yx\,dx$ could arise from the differentiation of some finite quantity. It is equally clear that the formula $P\,dx + Q\,dy$, if Q is equal to zero and P contains y, cannot be a real differential. At the same time we understand that the quantity Q cannot be chosen arbitrarily, but that it depends on P.

220. In order to investigate this relationship between P and Q in the differential $dV = P\,dx + Q\,dy$, we first suppose that V is a function of zero dimension in x and y. We proceed from particular cases to a general relation. Suppose that we let $y = tx$. Then the quantity x vanishes from the function V, and we have a function of t alone, which we call T, and whose differential is $\Theta\,dt$, where Θ is a function of t. We also substitute everywhere $y = tx$ and $dy = t\,dx + x\,dt$. From this we obtain $P\,dx + Qt\,dx + Qx\,dt$. Since dx is really not contained in this, we necessarily have $P + Qt = 0$, so that

$$Q = \frac{-P}{t} = \frac{-Px}{y}$$

and

$$Px + Qy = 0.$$

Hence in this case we have found the relation between P and Q. Furthermore, it is necessary that $\Theta = Qx$ so that Qx is equal to a function of t, that is, a function of zero dimension in x and y. Since $Q = \Theta/x$, we have $P = -\Theta y/x^2$, and both Px and Qy are functions of zero dimension in x and y.

221. If a function V is a function of zero dimension in x and y, and it is differentiated, then its differential $P\,dx + Q\,dy$ will always be such that $Px + Qy = 0$. That is, if in the differential instead of the differentials dx and dy we write x and y, the result will be equal to zero, as will appear from the following examples.

I. Let $V = \dfrac{x}{y}$. Then

$$dV = \frac{y\,dx - x\,dy}{y^2},$$

and when x replaces dx and y replaces dy, we have

$$\frac{yx - xy}{y^2} = 0.$$

II. Let $V = \dfrac{x}{\sqrt{x^2 - y^2}}$. Then

$$dV = \frac{-y^2 dx - yx\,dy}{\left(x^2 - y^2\right)^{3/2}},$$

so that

$$\frac{-y^2 x + y^2 x}{\left(x^2 - y^2\right)^{3/2}} = 0.$$

III. Let $V = \dfrac{y + \sqrt{x^2 + y^2}}{-y + \sqrt{x^2 + y^2}}$, which is a function of zero dimension in x and y. Then

$$dV = \frac{2x^2 dy - 2xy\,dx}{\left(\sqrt{x^2 + y^2} - y\right)^2 \sqrt{x^2 + y^2}},$$

and when x and y are substituted for dx and dy, the result is zero.

IV. Let $V = \ln\left(\dfrac{x + y}{x - y}\right)$. Then

$$dV = \frac{2x\,dy - 2y\,dx}{x^2 - y^2},$$

and

$$\frac{2xy - 2yx}{x^2 - y^2} = 0.$$

V. Let $V = \arcsin\left(\dfrac{\sqrt{x - y}}{\sqrt{x + y}}\right)$. Then

$$dV = \frac{y\,dx - x\,dy}{(x + y)\sqrt{2y}\,(x - y)},$$

and this formula enjoys the same property.

222. Now let us consider some other homogeneous functions and let V be an n-dimensional function of x and y. Hence if we let $y = tx$, then V takes the form Tx^n, where T is a function of t. We let $dT = \Theta\,dt$ so that

$$dV = x^n \Theta\,dt + nTx^{n-1}dx$$

If we let $dV = P\,dx + Q\,dy$, since $dy = t\,dx + x\,dt$, we have

$$dV = P\,dx + Qt\,dx + Qx\,dt.$$

Since the two forms should be equal, we have

$$P + Qt = nTx^{n-1} = \frac{nV}{x}$$

because $V = Tx^n$. Since $t = y/x$ we have

$$Px + Qy = nV,$$

and this equation thus defines the relation between P and Q. Hence if one is known, the other is easily discovered. Since $Qx = x^n\Theta$, it follows that Qx, as well as Qy and Px, is an n-dimensional function of x and y.

223. Hence, if in the differential of any homogeneous function in x and y we substitute x and y for dx and dy, respectively, the result will be equal to the original function whose differential is given multiplied by the dimension.

I. If $V = \sqrt{x^2 + y^2}$, then $n = 1$, and since

$$dV = \frac{x\,dx + y\,dy}{\sqrt{x^2 + y^2}},$$

we have

$$\frac{x^2 + y^2}{\sqrt{x^2 + y^2}} = V = \sqrt{x^2 + y^2}.$$

II. If $V = \dfrac{y^3 + x^3}{y - x}$, then $n = 2$ and

$$dV = \frac{2y^3dy - 3y^2x\,dy + 3yx^2dx - 2x^3dx + y^3dx - x^3dy}{(y-x)^2}.$$

When we substitute x for dx and y for dy, we have

$$\frac{2y^4 - 2y^3x + 2yx^3 - 2x^4}{(y-x)^2} = \frac{2y^3 + 2x^3}{y - x} = 2V.$$

III. If $V = \dfrac{1}{(y^2 + y^2)^2}$, then $n = -4$ and

$$dV = -\frac{4y\,dy + 4x\,dx}{(y^2 + x^2)^3}.$$

When we substitute x for dx and y for dy, we have

$$-\frac{4y^2 + 4x^2}{(x^2 + y^2)^3} = -4V.$$

IV. If $V = x^2 \ln\left(\frac{x+y}{y-x}\right)$, then $n = 2$ and

$$dV = 2x\,dx \ln\left(\frac{y+x}{y-x}\right) + \frac{2x^2\,(y\,dx - x\,dy)}{y^2 - x^2}.$$

When we make the required substitutions, there arises

$$2x^2 \ln\left(\frac{y+x}{y-x}\right) = 2V.$$

224. A similar property can be observed if V is a homogeneous function in more than two variables. If V is a function of the quantities x, y, and z, which together have a dimension of n, then its differential has the form $P\,dx + Q\,dy + R\,dz$. When we let $y = tx$ and $z = sx$, so that $dy = t\,dx + x\,dt$ and $dz = s\,dx + x\,ds$, then furthermore, the function V takes the form Ux^n, where U is a function of the two variables t and s. Hence, if we let $dU = p\,dt + q\,ds$, then

$$dV = x^n pdt + x^n q\,ds + nUx^{n-1}dx.$$

The previous form gives

$$dV = P\,dx + Qt\,dx + Qx\,dt + Rs\,dx + Rx\,ds.$$

When these two forms are compared, we have

$$P + Qt + Rs = nUx^{n-1} = \frac{nV}{x},$$

from which we obtain

$$Px + Qy + Rz = nV.$$

This same property extends to functions of howsoever many variables.

225. If the given function is homogeneous in howsoever many variables $x, y, z, v, \ldots$, its differential will always have the property that, if for the differentials $dx, dy, dz, dv, \ldots$ we substitute the finite quantities $x, y, z, v, \ldots$, then the result is the given function multiplied by the dimension. This rule applies likewise if V is a homogeneous function of the single variable x. In this case V is a power of x, for example $V = ax^n$, which is a homogeneous function of dimension n. Indeed, there is no other function of x in which x has n dimensions besides the power x^n. Since $dV = nax^{n-1}dx$, when we substitute x for dx, we obtain nax^n, which is nV. This remarkable property of homogeneous functions deserves to be very carefully noted, since it will have extremely useful consequences in integral calculus.

226. Now, in order to inquire into the general relationship between P and Q, which constitute the differential $P\,dx + Q\,dy$ of any function V of two variables x and y, we need to pay attention to what follows. If V is any function whatsoever of x and y, and we substitute $x + dx$ for x, then V is transformed into R. If $y + dy$ is substituted for y, then V is transformed into S. If simultaneously $x+dx$ and $y+dy$ are substituted for x and y, then V is changed into V^{I}. Since R comes from V when $x + dx$ is substituted for x, it is clear that if furthermore $y + dy$ is substituted in R, the result is V^{I}. It comes to the same thing as substituting $x + dx$ for x and $y + dy$ for y immediately. In a similar way, if $x + dx$ is substituted for x in S, since S has already arisen from V by substituting $y + dy$ for y, once again we obtain V^{I}, as may be seen more clearly from the following table.

The quantity	becomes	if for	we put
V	R	x	$x + dx$
V	S	y	$y + dy$
V	V^{I}	x	$x + dx$
		y	$y + dy$
R	V^{I}	y	$y + dy$
S	V^{I}	x	$x + dx$

227. If we differentiate V as if x were the only variable and y is treated as a constant, since we substitute $x+dx$ for x, the function V becomes R, whose differential will be equal to $R-V$. From the form $dV = P\,dx+Q\,dy$ it follows that the same differential will be equal to $P\,dx$, so that $R-V = P\,dx$. If we substitute $y + dy$ for y and treat x as a constant, since R becomes V^{I} and V becomes S; the quantity $R - V$ becomes $V^{\mathrm{I}} - S$. Then the differential of $R - V = P\,dx$, which arises if only y is considered variable, will be equal to

$$V^{\mathrm{I}} - R - S + V.$$

In a similar way, when we substitute $y+dy$ for y, V becomes S, so that $S-V$ is the differential of V if we let only y be variable, so that $Q - V = Q\,dy$. Now when we substitute $x + dx$ for x, S becomes V^{I} and V becomes R, so that the quantity $S-V$ becomes $V^{\mathrm{I}}-R$ and the differential of $S-V = Q\,dy$, which arises when only x is variable, is equal to

$$V^{\mathrm{I}} - R - S + V,$$

which is equal to the differential we found previously.

228. From this equality we deduce the following conclusion. If any function V of two variables x and y has a differential $dV = P\,dx + Q\,dy$, then the differential of $P\,dx$, which comes from letting only y be variable while x is held constant, is equal to the differential of $Q\,dy$, which comes from letting only x be variable while y is held constant. For instance, if only y is variable, then $dP = Z\,dy$ and the differential of $P\,dx$, taken in the prescribed way, will be equal to $Z\,dx\,dy$. Now, if we let only x be variable, then also $dQ = Z\,dx$. Thus the differential of $Q\,dy$, taken in the prescribed way, will be $Z\,dx\,dy$. In this way we understand the relationship between P and Q. In short, the differential of $P\,dx$ when x is constant must be equal to the differential of $Q\,dy$ when y is constant.

229. This remarkable property will become clearer if we illustrate it with a few examples.

I. Let $V = yx$. Then

$$dV = y\,dx + x\,dy,$$

so that $P = y$ and $Q = x$. When we keep x constant,

$$d.P\,dx = dx\,dy,$$

and when y is kept constant,

$$d.Q\,dy = dx\,dy,$$

so that the two differentials are equal.

II. Let $V = \sqrt{x^2 - 2xy}$. Then

$$dV = \frac{x\,dx + y\,dx + x\,dy}{\sqrt{x^2 + 2xy}},$$

so that

$$P = \frac{x + y}{\sqrt{x^2 + 2xy}} \quad \text{and} \quad Q = \frac{x}{\sqrt{x^2 + 2xy}},$$

so that when x is kept constant,

$$d.P\,dx = \frac{xy\,dx\,dy}{(x^2 + 2xy)^{3/2}},$$

and when y is kept constant,

$$d.Q = \frac{xy\,dx\,dy}{(x^2 + 2xy)^{3/2}}.$$

III. Let $V = x \sin y + y \sin x$, so that

$$dV = dx \sin y + x\, dy \cos y + dy \sin x + y\, dx \cos x.$$

Hence $P\, dx = dx \sin y + y\, dx \cos x$ and $Q\, dy = dy \sin x + x\, dy \cos y$. When we keep x constant, we have

$$d.P\, dx = dx\, dy \cos y + dx\, dy \cos x,$$

and when y is kept constant, we have

$$d.Q\, dy = dx\, dy \cos y + dx\, dy \cos x.$$

IV. Let $V = x^y$. Then

$$dV = x^y\, dy \ln x + yx^{y-1} dx,$$

so that $P\, dx = yx^{y-1} dx$ and $Q\, dy = x^y dy \ln x$. Hence when we keep x constant we have

$$d.P\, dx = x^{y-1} dx\, dy + yx^{y-1} dx\, dy \ln x,$$

and when y is kept constant, we have

$$d.Q\, dy = yx^{y-1} dx\, dy \ln x + x^{y-1} dx\, dy.$$

230. This property can also be stated in another way, so that this remarkable characteristic of all functions of two variables can be understood. If any function V of two variables x and y is differentiated with only x variable, and this differential is again differentiated with only y variable, then after this double differentiation, the same result is obtained when the order of differentiation is reversed by first differentiating V with only y variable and then differentiating this differential with only x variable. Both of these cases give the same expression of the form $z\, dx\, dy$. The reason for this identity clearly follows from the previous property; for if V is differentiated with only x variable, we have $P\, dx$, and if V is differentiated with only y variable, we have $Q\, dy$. The differentials of these, in the way already indicated, are equal, as we have demonstrated. For the rest, this characteristic follows immediately from the argument given in paragraph 227.

231. The relationship between P and Q, if $P\, dx + Q\, dy$ is the differential of the function V, can also be indicated in the following way. Since P and Q are functions of x and y, they can both be differentiated with both x and y variable. If $dV = P\, dx + Q\, dy$, then $dP = p\, dx + r\, dy$ and $dQ = q\, dx + s\, dy$. Therefore, when x is constant, $dP = r\, dy$ and $d.P\, dx = r\, dx\, dy$. Then when y is constant $dQ = q\, dx$ and $d.Q\, dy = q\, dx\, dy$. Since these two differentials $r\, dx\, dy$ and $q\, dx\, dy$ are equal to each other, it follows that

$$q = r.$$

Therefore, the functions P and Q are related in such a way that if both of them are differentiated as we have done, the quantities q and r are equal to each other. For the sake of brevity, at least in this chapter, the quantities r and q will conveniently be symbolized by $\partial P/\partial y$, in that P is differentiated with only y variable, which is indicated by ∂y in the denominator. In this way we obtain the finite quantity r. In like manner $\partial Q/\partial x$ will symbolize the finite quantity q, since by this is indicated that the function Q is differentiated with only x variable, so we ought to divide the differential by ∂x.

232. We will use this method of symbolizing, although there is some danger of ambiguity therein. We avoid this with the use of the partial symbol, with the result that the complications in describing the conditions of differentiation are avoided. Thus we can briefly express the relationship between P and Q by stating

$$\frac{\partial P}{\partial y} = \frac{\partial Q}{\partial x}.$$

In fractions of this kind, beyond the usual significance is which the denominator indicates the divisor, here the differential of the numerator is to be taken with only that quantity variable which is indicted by the differential in the denominator. In this way by the division of differentials these fractions $\partial P/\partial y$ and $\partial Q/\partial x$ exist from calculus and indicate finite quantities, which in this case are equal to each other. Once this method is agreed upon, the quantities p and s can be denoted by $p = \partial P/\partial x$ and $s = \partial Q/\partial y$ if, as we have noted, the differentiation of the numerator by the denominator is properly restricted.

233. There is a wonderful agreement between this property and that property of homogeneous functions which we previously have shown. Let V be a homogeneous function in x and y of dimension n. We let $dV = P\,dx + Q\,dy$, and we have shown that $nV = Px + Qy$. Hence

$$Q = \frac{nV}{y} - \frac{Px}{y}.$$

We let $dP = p\,dx + r\,dy$, so that

$$\frac{\partial P}{\partial y} = r,$$

which is thus shown to be equal to $\partial Q/\partial x$. Let Q be differentiated with only x variable, and under the present hypothesis we have

$$dQ = \frac{nP\,dx}{y} - \frac{P\,dx}{y} - \frac{xp\,dx}{y},$$

so that

$$\frac{\partial Q}{\partial x} = \frac{(n-1)\,P}{y} - \frac{px}{y},$$

and we must have

$$\frac{(n-1)\,P}{y} - \frac{px}{y} = r,$$

or

$$(n-1)\,P = px + ry.$$

This equality becomes clear when we note that P is a homogeneous function in x and y of dimension $n-1$, so that its differential $dP = p\,dx + r\,dy$, due to the property of homogeneous functions, should be such that $(n-1)\,P = px + ry$.

234. This property, that

$$\frac{\partial P}{\partial y} = \frac{\partial Q}{\partial x},$$

which we have shown to be common to all functions of two variables x and y, can also reveal to us the nature of functions of three or more variables. Let V be any function of three variables x, y, and z, and let

$$dV = P\,dx + Q\,dy + R\,dz.$$

If in this differentiation z is treated as a constant, then $dV = P\,dx + Q\,dy$. In this case by what has gone before it should be true that $\partial P/\partial y = \partial Q/\partial x$. Then if y is supposed to be constant, we have $dV = P\,dx + R\,dz$, so that $\partial P/\partial z = \partial R/\partial x$. Finally with x constant we see that $\partial Q/\partial z = \partial R/\partial y$. Therefore, in the differential $P\,dx + Q\,dy + R\,dz$ of the function V, the quantities P, Q, and R are related to each other in such a way that

$$\frac{\partial P}{\partial y} = \frac{\partial Q}{\partial x}, \qquad \frac{\partial P}{\partial z} = \frac{\partial R}{\partial x}, \qquad \frac{\partial Q}{\partial z} = \frac{\partial R}{\partial y}.$$

235. It follows that this property of functions that involve three or more variables is analogous to that which we have shown above (paragraph 230) for functions of two variables. If V is any function of three variables x, y, and z, and this is differentiated three times in such a way that in the first differentiation the first variable, that is, x, is the only variable, in the second differentiation only y is variable, and in the third only z is variable, then we obtain an expression of the form $Z\,dx\,dy\,dz$. This same expression is obtained no matter in which order the quantities x, y, and z are taken. There are six different ways of taking the threefold derivative to obtain the same expression, since there are six ways of ordering x, y, and z. No matter what order is chosen, if the function V is differentiated with only the first variable, and that is then differentiated with only the second variable, and this then differentiated with only the third variable, the same expression is obtained when the order is changed.

236. In order that the reason for this property may be seen more clearly, we let

$$dV = P\,dx + Q\,dy + R\,dz.$$

Then we differentiate each of the quantities P, Q, and R, with their differentials of the form we have already seen:

$$\begin{aligned} dP &= p\,dx + s\,dy + t\,dz, \\ dQ &= s\,dx + q\,dy + u\,dz, \\ dR &= t\,dx + u\,dy + r\,dz. \end{aligned}$$

Now if we differentiate V with only x variable, we have $P\,dx$. This differential is now differentiated with only y variable to obtain $s\,dx\,dy$. If this is differentiated with only z variable, and after this is divided by $dx\,dy\,dz$, we have $\partial s/\partial z$. Now we reorder the variables as y, x, z, and the first differentiation gives $Q\,dy$, the second $s\,dx\,dy$ and the third (when divided by $dx\,dy\,dz$) gives $\partial s/\partial z$ as before. Now choose the order z, y, x and the first differentiation gives $R\,dz$, the second $u\,dy\,dz$, and the third, after division by $dx\,dy\,dz$, gives $\partial u/\partial x$. But when y is kept constant, we have $dQ = s\,dx + u\,dz$, so that

$$\frac{\partial s}{\partial z} = \frac{\partial u}{\partial x},$$

as we wished to show.

237. We let

$$V = \frac{x^2 y}{a^2 - z^2},$$

and we take the three derivatives as many times as the order of the variables x, y, and z can change:

	1st differential	2nd differential	3rd differential
with respect to	x alone	y alone	z alone
	$\frac{2xy\,dx}{a^2-z^2}$	$\frac{2x\,dx\,dy}{a^2-z^2}$	$\frac{4xz\,dx\,dy\,dz}{(a^2-z^2)^2}$
with respect to	x alone	z alone	y alone
	$\frac{2xy\,dx}{a^2-z^2}$	$\frac{4xyz\,dx\,dz}{(a^2-z^2)^2}$	$\frac{4xz\,dx\,dy\,dz}{(a^2-z^2)^2}$
with respect to	y alone	x alone	z alone
	$\frac{x^2dy}{a^2-z^2}$	$\frac{2x\,dx\,dy}{a^2-z^2}$	$\frac{4xz\,dx\,dy\,dz}{(a^2-z^2)^2}$
with respect to	y alone	z alone	x alone
	$\frac{x^2dy}{a^2-z^2}$	$\frac{2x^2z\,dy\,dz}{(a^2-z^2)^2}$	$\frac{4xz\,dx\,dy\,dz}{(a^2-z^2)^2}$
with respect to	z alone	x alone	y alone
	$\frac{2x^2yz\,dz}{(a^2-z^2)^2}$	$\frac{4xyz\,dx\,dz}{(a^2-z^2)^2}$	$\frac{4xz\,dx\,dy\,dz}{(a^2-z^2)^2}$
with respect to	z alone	y alone	x alone
	$\frac{2x^2yz\,dz}{(a^2-z^2)^2}$	$\frac{2x^2z\,dy\,dz}{(a^2-z^2)^2}$	$\frac{4xz\,dx\,dy\,dz}{(a^2-z^2)^2}$

From this example it is clear that no matter in what order the variables are taken, after the three differentiations we always have the same expression

$$\frac{4xz\,dx\,dy\,dz}{(a^2-z^2)^2}.$$

238. Just as after three differentiations we arrived at the same expression, so we detect some agreement after the second differentiation. Among these each expression occurs twice. It is clear that those formulas with the same differentials are equal to each other, and the third differentials are all equal to each other because they all have the same differentials $dx\,dy\,dz$. From this we conclude that if V is a function of howsoever many variables $x, y, z, v, u, \ldots$ and V is differentiated successively the number of times required so that always only one quantity is variable, then as often as we arrive at expressions with the same differentials, those expressions will be equal to each other. Thus, after two differentiations we find an expression $Z\,dx\,dy$ where in one only x is variable and in the other only y is variable, no matter which is first or second. In a similar way there are six ways by which the triple differentiation arrives at the same expression $Z\,dx\,dy\,dz$. There are twenty-four ways of taking four derivatives to arrive at the same expression of the form $Z\,dx\,dy\,dz\,dv$, and so forth.

239. One can easily agree to the truths of these theorems with little attention paid to the principles explained earlier, and one can more easily see this truth by one's own meditation than by such complications of words, without which it is not possible to give a demonstration. But since a knowledge of these properties is of the greatest importance in integral calculus, beginners should be warned that they must not only meditate on these properties with great care and examine their truth, but also work through many examples. In this way they will become very familiar with this material, and then they will be able more easily to gather the fruit which will come later. Indeed, not only beginners, but also those who are already acquainted with the principles of differential calculus are exhorted to the same, since in almost all introductions to this part of analysis this argument is wont to be omitted. Frequently, authors have been content to give only the rules for differentiation and the applications to higher geometry. They do not inquire into the nature or the properties of differentials, from which the greatest aid to integral calculus comes. For this reason the argument, which is practically new in this chapter, has been discussed at length in order that the way to other more difficult integrations may be prepared, and the work to be undertaken later might be lightened.

240. Once we know these properties that functions of two or more variables enjoy, we can easily decide whether or not a given formula for a differential in which there occur two or more variables has really arisen from differentiation of some finite function. If in the formula $P\,dx + Q\,dy$ it is not true that $\partial P/\partial y = \partial Q/\partial x$, then we can with certainty state that there is no function of x and y whose differential is equal to $P\,dx + Q\,dy$. When we come to integral calculus we will deny that there is any integral for such a formula. Hence since $yx\,dx + x^2dy$ does not have the required condition, there is no function whose differential is equal to $yx\,dx + x^2dy$. The question is whether as long as $\partial P/\partial y = \partial Q/\partial x$, the formula has always arisen from the differentiation of some function. From the principles of integration we can surely answer in the affirmative.

241. If in a given formula of a differential there are three or more variables, such as $P\,dx + Q\,dy + R\,dz$, then there is no way that this shall have arisen from differentiation unless these three conditions are met:

$$\frac{\partial P}{\partial y} = \frac{\partial Q}{\partial x}, \qquad \frac{\partial P}{\partial z} = \frac{\partial R}{\partial x}, \qquad \frac{\partial Q}{\partial z} = \frac{\partial R}{\partial y}.$$

Of these conditions, if even only one is missing, then we can state with certainty that there is no function of x, y, and z whose differential is $P\,dx + Q\,dy + R\,dz$. For such a formula we cannot require an integral, and such is said not to be integrable. It can easily be understood that in integral calculus we ought to investigate the differential formulas to see whether they can be integrated before actually working to find the integral.

8

On the Higher Differentiation of Differential Formulas

242. If there is a single variable and its differential is held constant, we have already given the method for finding differentials of any order. That is, if the differential of any function is differentiated again, we obtain its second differential. If this is again differentiated, we get the third differential, and so forth. This same rule holds whether the function involves several variables or only one, whose first differential is not kept constant. Hence, if V is any function of x and dx is not held constant, but is as if it were a variable, then the differential of dx is equal to d^2x. The differential of d^2x is equal to d^3x, and so forth. Let us investigate the second and higher differentials of the function V.

243. We let the first differential of the function V be equal to $P\,dx$, where P is some function of x depending on V. If we want to find the second differential of V, we must differentiate again its first differential $P\,dx$. Since this is the product of two variable quantities, P and dx, whose differentials respectively are $dP = p\,dx$ and $d.dx = d^2x$, then according to the product rule the second differential is

$$d^2V = P\,d^2x + p\,dx^2.$$

Then if we let $dp = q\,dx$, since the differential of dx^2 is equal to $2dx\,d^2x$, we have by another differentiation

$$d^3V = P\,d^3x + dP\,d^2x + 2p\,dx\,d^2x + dp\,dx^2.$$

Since $dP = p\,dx$ and $dp = q\,dx$, we have

$$d^3V = P\,d^3x + 3p\,dx\,d^2x + q\,dx^3.$$

We find the higher differentials in a similar way.

244. Now we apply this to powers of x, whose successive differentials we investigate, supposing that dx is not kept constant.

I. If $V = x$, then

$$dV = dx, \qquad d^2V = d^2x, \qquad d^3V = d^3x, \qquad d^4V = d^4x, \qquad \ldots .$$

II. If $V = x^2$, then

$$dV = 2x\,dx$$

and

$$\begin{aligned} d^2V &= 2x\,d^2x + 2dx^2, \\ d^3V &= 2x\,d^3x + 6dx\,d^2x, \\ d^4V &= 2x\,d^4x + 8dx\,d^3x + 6d^2x^2, \\ d^5V &= 2x\,d^5x + 10dx\,d^4x + 20d^2x\,d^3x, \\ &\ldots . \end{aligned}$$

III. If in general $V = x^n$, then and

$$\begin{aligned} dV &= nx^{n-1}dx, \\ d^2V &= nx^{n-1}d^2x + n(n-1)\,x^{n-2}dx^2, \\ d^3V &= nx^{n-1}d^3x \\ &\quad + 3n(n-1)\,x^{n-2}dx\,d^2x + n(n-1)(n-2)\,x^{n-3}dx^3, \\ d^4V &= nx^{n-1}d^4x + 4n(n-1)\,x^{n-2}dx\,d^3x + 3n(n-1)\,x^{n-2}d^2x^2 \\ &\quad + 6n(n-1)(n-2)\,x^{n-3}dx^2d^2x \\ &\quad + n(n-1)(n-2)(n-3)\,x^{n-4}dx^4, \\ &\ldots . \end{aligned}$$

If dx happens to be constant, then $d^2x = 0$, $d^3x = 0$, $d^4x = 0$, and so forth. Thus we have the same differentials we found earlier under this hypothesis.

245. Since differentials of any order are differentiated according to the same rules as are finite quantities, any expression in which differentials occur besides finite quantities can be differentiated according to the rules given above. Since this operation occurs not infrequently, we illustrate this with a few examples.

I. If $V = \dfrac{x\,d^2x}{dx^2}$, we differentiate to obtain

$$dV = \frac{x\,d^3x}{dx^2} + \frac{d^2x}{dx} - \frac{2x\,d^2x^2}{dx^3}.$$

II. If $V = \dfrac{x}{dx}$, then

$$dV = 1 - \frac{x\,d^2x}{dx^2},$$

where there is no problem if we let V be an infinite quantity.

III. If

$$V = x^2 \ln\left(\frac{d^2x}{dx^2}\right),$$

we first transform V into $x^2 \ln d^2x - 2x^2 \ln dx$, and then by the ordinary rules for differentiating we have

$$dV = 2x\,dx \ln d^2x + \frac{x^2d^3x}{d^2x} - 4x\,dx \ln dx - \frac{2x^2d^2x}{dx}.$$

The higher differentials of V can be found in a similar way.

246. If the given expression involves two variables, namely x and y, either only one of the differentials can be kept constant, or neither. It is arbitrary which of the differentials is assumed to be constant, since it depends on our choice of the extent to which we want successive values to increase. However, we cannot decide to keep both differentials constant, since this would assume some relationship between x and y, while we have assumed that there is no such relationship, or if there is, that it is unknown. If we suppose that x increases equally and y also takes equal increments, then by that fact we would have $y = ax + b$, and hence y would depend on x, which is contrary to the hypothesis. For this reason either only one differential of a variable is kept constant, or neither is kept constant. If we know how to perform differentiations with no differential taken as constant, it is clear how to find differentials if one differential is kept constant: If dx is constant, we need only let the terms that contain d^2x, d^3x, d^4x, and so forth, be deleted.

247. We let V be any finite function of x and y, and let $dV = P\,dx + Q\,dy$. In order to find the second differential of V we suppose that both of the differentials dx and dy are variable. Since P and Q are functions of x and y, we let

$$dP = p\,dx + r\,dy,$$

$$dQ = r\,dx + q\,dy,$$

since we have already noted that

$$\frac{\partial P}{\partial y} = \frac{\partial Q}{\partial x} = r.$$

Under these conditions we differentiate $dV = P\,dx + Q\,dy$ and obtain

$$d^2V = P\,d^2x + p\,dx^2 + 2r\,dx\,dy + Q\,d^2y + q\,dy^2.$$

Hence, if we suppose that dx is constant, then

$$d^2V = p\,dx^2 + 2r\,dx\,dy + Q\,d^2y + q\,dy^2.$$

On the other hand, if we suppose that dy is constant, then

$$d^2V = P\,d^2x + p\,dx^2 + 2r\,dx\,dy + q\,dy^2.$$

248. Therefore, if any function of x and y is differentiated twice, with neither differential held constant, the second differential always has the form

$$d^2V = P\,d^2x + Q\,d^2y + R\,dx^2 + S\,dy^2 + T\,dx\,dy;$$

where the quantities P, Q, R, S, and T are so interrelated that when we use the notation used in the previous chapter,

$$\frac{\partial P}{\partial y} = \frac{\partial Q}{\partial x}, \qquad R = \frac{\partial P}{\partial x}, \qquad S = \frac{\partial Q}{\partial y}, \qquad T = 2\frac{\partial Q}{\partial x} = 2\frac{\partial P}{\partial y}.$$

If any one of these conditions fails, then we can affirm with certainty that the proposed formula cannot be the second differential of a function. Here we have an immediate test of whether or not an expression of this kind is the second differential of some quantity.

249. In a similar way the third differential and higher differentials are found. It would seem to be helpful to show a particular example rather than give general formulas.

Let $V = xy$, so that

$$\begin{aligned} dV &= y\,dx + x\,dy, \\ d^2V &= y\,d^2x + 2dx\,dy + x\,d^2y, \\ d^3V &= y\,d^3x + 3dy\,d^2x + 3d^2y\,dx + x\,d^3y, \\ d^4V &= y\,d^4x + 4dy\,d^3x + 6d^2x\,d^2y + 4dx\,d^3y + x\,d^4y, \\ &\ldots. \end{aligned}$$

In this example the numerical coefficients follow the law of the powers of a binomial, so that this can be continued howsoever far one wishes.

If $V = y/x$, then

$$\begin{aligned} dV &= \frac{dy}{x} - \frac{y\,dx}{x^2}, \\ d^2V &= \frac{d^2y}{x} - \frac{2dx\,dy}{x^2} + \frac{2y\,dx^2}{x^3} - \frac{y\,d^2x}{x^2}, \\ d^3V &= \frac{d^3y}{x} - \frac{3dx\,d^2y}{x^2} + \frac{6dx^2dy}{x^3} - \frac{3dy\,d^2x}{x^2} \\ &\quad + \frac{6y\,dx\,d^2x}{x^3} - \frac{6y\,dx^3}{x^4} - \frac{y\,d^3x}{x^2}, \\ &\ldots. \end{aligned}$$

In this example the sequence of differentials is not as clear as in the previous example.

250. This method of differentiation is not confined only to finite functions. It can also be extended to any expression that already contains differentials. The differential can be found whether or not some differential is assumed to remain constant. Since each differential is differentiated by the same laws as finite quantities, the rules given in the preceding chapters are still valid and should be observed. Let V denote such an expression that we need to differentiate, whether it is finite or infinitely large or infinitely small. The method of differentiation can be seen from the following examples.

I. Let $V = \sqrt{dx^2 + dy^2}$. Then

$$dV = \frac{dx\,d^2x + dy\,d^2y}{\sqrt{dx^2 + dy^2}}.$$

II. Let $V = \dfrac{y\,dx}{dy}$. Then

$$dV = dx + \frac{y\,d^2x}{dy} - \frac{y\,dx\,d^2y}{dy^2}.$$

III. Let

$$V = \frac{\left(dx^2 + dy^2\right)^{3/2}}{dx\, d^2y - dy\, d^2x}.$$

Then

$$dV = \frac{\left(3dx\, d^2x + 3dy\, d^2y\right)\sqrt{dx^2 + dy^2}}{dx\, d^2y - dy\, d^2x} - \frac{\left(dx^2 + dy^2\right)^{3/2}\left(dx\, d^3y - dy\, d^3x\right)}{\left(dx\, d^2y - dy\, d^2x\right)^2}.$$

Since these differentials are taken most generally, with no differential taken to be constant, from these it is easy to derive the differentials that arise when either dx or dy is held constant.

251. Since we are assuming that none of the differentials are constant, we can give no law according to which the second differentials and those of higher order can be determined, nor do they have a definite meaning. Hence the formula for the second differential and those of higher order have no determined value, unless some differential is assumed to be constant. But even its signification will be vague and will change depending on which of the differentials are held constant. There are, however, some expressions that for second differentials, although no differential is held constant, still have a determined signification. This always remains the same, no matter which differential we decide to hold constant. Below we will consider very carefully the nature of formulas of this kind, and we will discuss the way in which these may be distinguished from those others that do not include any determined values.

252. In order that we may more easily see the kind of formulas that contain second or higher differentials, we consider first formulas containing only a single variable. It will then be perfectly clear that if in such a formula there is a second differential of the variable x, d^2x, and no differential is held constant, then it is not possible for the formula to have a fixed value. Indeed, if we decided that the differential of x should be constant, then $d^2x = 0$. However, if we held constant the differential of x^2, that is, $2x\, dx$, or even $x\, dx$, since the differential of $x\, dx$ is $x\, d^2x + dx^2$, this expression is equal to zero, so that $d^2x = -dx^2/x$. Indeed, if the differential of some power, for example $nx^{n-1}dx$ or $x^{n-1}dx$, should be constant, then its second differential satisfies

$$x^{n-1}d^2x + (n-1)\, x^{n-2}dx^2 = 0,$$

so that

$$d^2x = -\frac{(n-1)\, dx^2}{x}.$$

Different values for d^2x will be given if the differentials of other functions of x are held constant. However, it is clear that the formulas in which d^2x occurs take on quite different values depending on whether in place of d^2x we write zero or $-dx^2/x$ or $-(n-1)\,dx^2/x$ or some other expression of this kind. For instance, if the given formula is x^2d^2x/dx^2, then, because d^2x and dx^2 are both infinitely small and homogeneous, the expression should have a finite value. If dx is made constant, the expression becomes zero; if $d.x^2$ is constant, it becomes $-x$; if $d.x^3$ is constant, it becomes $-2x$; if $d.x^4$ is constant, it becomes $-3x$, and so forth. Hence, it can have no determined value unless the differential of something is assumed to be constant.

253. This ambiguity of signification is present, for a similar reason, if the third differential is present in some formula. Let us consider the formula

$$\frac{x^3d^3x}{dx\,d^2x},$$

which also has a finite value. If the differential dx is constant, then the formula takes the form $0/0$, whose value we will soon see. Let $d.x^2$ be constant. Then $d^2x = -dx^2/x$ and after another differentiation we obtain

$$d^3x = -\frac{2dx\,d^2x}{x} + \frac{dx^3}{x^2} = \frac{3dx^3}{x^2},$$

since $d^2x = -dx^2/x$. Hence, for this reason, the given formula

$$\frac{x^3d^3x}{dx\,d^2x}$$

becomes $-3x^2$. However, if $d.x^n$ is constant, then

$$d^2x = \frac{-(n-1)\,dx^2}{x},$$

so that

$$d^3x = -\frac{2(n-1)\,dx\,d^2x}{x} + \frac{(n-1)\,dx^3}{x^2} = \frac{2(n-1)^2\,dx^3}{x^2} + \frac{(n-1)\,dx^3}{x^2},$$

$$= \frac{(2n-1)(n-1)\,dx^3}{x^2}.$$

Hence for this reason we have

$$\frac{d^3x}{d^2x} = \frac{(2n-1)\,dx}{x}$$

and

$$\frac{x^3d^3x}{dx\,d^2x} = -(2n-1)\,x^2.$$

It follows that if $n = 1$, or dx is constant, the value of the formula will be equal to $-x^2$. From this it is clear that if in any formula there occurs a third or higher differential and at the same time it is not indicated which of these differentials are taken to be constant, then that formula has no certain value and can have no further significance. For this reason such expressions cannot occur in the calculation.

254. In a similar way, if the formula contains two or more variables and there occur differentials of the second or higher order, it should be understood that it can have no determined value unless some differential is constant, with the exception of some special cases that we will soon consider. Since as soon as d^2x is in some formula, due to the various differentials that can be constant, the value of d^2x always changes. The result is that it is impossible that the formula should have a stated value. The same is true for any higher differential of x and also for the second and higher differentials of the other variables. However, if a formula contains the differentials of two or more variables, it can happen that the variability arising from one is destroyed by the variability of the others. It is for this reason that we have that exceptional case that we mentioned, in which a formula of this kind, involving second differentials of two or more variables, can have a definite value, even though no differential is held constant.

255. The formula

$$\frac{y\,d^2x + x\,d^2y}{dx\,dy}$$

can have no fixed and stated signification unless one of the first differentials is set constant. If dx is made constant, then we have

$$\frac{x\,d^2y}{dx\,dy}.$$

On the other hand, if dy is made constant, we have

$$\frac{y\,d^2x}{dx\,dy}.$$

It should be clear that these formulas need not be equal. If they were necessarily equal, they would remain the same when any function of x is substituted for y. Let us suppose that $y = x^2$. When we set dx constant we have $d^2y = 2dx^2$, and the formula

$$\frac{x\,d^2y}{dx\,dy}$$

becomes equal to 1. However, if dy, that is $2x\,dx$, is set constant, then $d^2y = 2x\,d^2x + 2dx^2 = 0$, so that $d^2x = -dx^2/x$, and the formula

$$\frac{y\,d^2x}{dx\,dy}$$

becomes equal to $-\frac{1}{2}$. Since we have this contradiction in a single case, much less is it possible in general that

$$\frac{x\,d^2y}{dx\,dy},$$

when dx is constant, is equal to

$$\frac{y\,d^2x}{dx\,dy},$$

when dy is constant. Since the formula

$$\frac{y\,d^2x + x\,d^2y}{dx\,dy}$$

has no fixed meaning even though either dx or dy is constant, much less will there be a fixed meaning if the differential of an arbitrary function of either x of y or both is set equal to a constant.

256. Thus it appears that a formula of this kind cannot have a stated value unless it is so made up that when for y or z or any function of x is substituted, the second and higher differentials of x, namely d^2x, d^3x, etc., no longer remain in the calculation. Indeed, if after any such substitution whatsoever in the formula there remains d^2x or d^3x or d^4x, etc., the value of this formula remains unsettled. This is because as different constants are assigned, the differentials take on different meanings. The formula we have just discussed,

$$\frac{y\,d^2x + x\,d^2y}{dx\,dy},$$

is of this kind. If this formula had a fixed value, no matter what y should signify, the stated value should remain the same if y represents any function or x. But if we let $y = x$, the formula becomes $2x\,d^2x/dx^2$, which is undetermined due to the presence of d^2x, so that it takes on various values according to the various differentials that are made constant. This should be sufficiently clear from the discussion in paragraph 252.

257. From this there arises a doubt as to the existence of any formulas that contain two or more second, or higher, differentials that still have the property that when arbitrary functions of one of the variables are substituted for the other variables, then the second differentials are eliminated. We propose this doubt in order to present a formula that has this precise property. By this investigation we will more easily see the force of the question. I say that the following formula has this remarkable property:

$$\frac{dy\,d^2x + dx\,d^2y}{dx^3}.$$

Indeed, no matter what function of x we substitute for y, the second differentials always vanish completely.

I. Let $y = x^2$. Then $dy = 2xdx$ and $d^2y = 2x\,d^2x + 2dx^2$. When these values are substituted into the formula

$$\frac{dy\,d^2x - dx\,d^2y}{dx^3},$$

we have

$$\frac{2x\,dx\,d^2x - 2x\,dx\,d^2x - 2dx^3}{dx^3} = -2.$$

II. Let $y = x^n$. Then $dy = nx^{n-1}dx$ and

$$d^2y = nx^{n-1}d^2x + n\,(n-1)\,x^{n-2}dx^2.$$

When these values are substituted, the formula

$$\frac{dy\,d^2x - dx\,d^2y}{dx^3}$$

is transformed into

$$\frac{nx^{n-1}dx\,d^2x - nx^{n-1}dx\,d^2x - n\,(n-1)\,x^{n-2}dx^3}{dx^3} = -n\,(n-1)\,x^{n-2}.$$

III. Let $y = -\sqrt{1-x^2}$. Then

$$dy = \frac{x\,dx}{\sqrt{1-x^2}}$$

and

$$d^2y = \frac{x\,d^2x}{\sqrt{1-x^2}} + \frac{dx^2}{(1-x^2)^{3/2}},$$

so that the formula

$$\frac{dy\,d^2x - dx\,d^2y}{dx^3}$$

becomes

$$\frac{x\,d^2x}{dx^2\sqrt{1-x^2}} - \frac{x\,d^2x}{dx^2\sqrt{1-x^2}} - \frac{1}{(1-x^2)^{3/2}} = \frac{-1}{(1-x^2)^{3/2}}.$$

In all of these examples the second differentials d^2x cancel each other. This also happens no matter what other functions of x are substituted for y.

258. Since these examples have already shown the truth of our proposition, namely, that the formula

$$\frac{dy\,d^2x - dx\,d^2y}{dx^3}$$

has a fixed value, even though no differential is assumed to be constant, all the more easily we can furnish a demonstration. Let y be any function or x, and then dy is its differential, so that $dy = p\,dx$, where p is some function of x. The differential of p will have the form $dp = q\,dx$ and q is a function of x. Since $dy = p\,dx$, by differentiation we have $d^2y = pd^2x + q\,dx^2$ and

$$dy\,d^2x - dx\,d^2y = p\,dx\,d^2x - p\,dx\,d^2x - q\,dx^3 = -q\,dx^3.$$

In this expression, since there is no second differential, it has a fixed value and

$$\frac{dy\,d^2x - dx\,d^2y}{dx^3} = -q.$$

No matter how y depends on x, the second differentials in this formula vanish. For this reason its value is quite fixed, although in other respects it may be unsettled.

259. Although we have here supposed that y is a function of x, nevertheless the truth of the assertion remains true even if y does not depend on x at all. While we substitute for y an arbitrary function, and whatever kind it might be we do not determine, we attribute to y no dependence on x. Meanwhile, with no mention of a function, a demonstration can be given. No matter what quantity y might be, whether it depends on x or not, its differential dy will be homogeneous with dx, so that dy/dx will be some finite quantity p. The differential of p that we take when x goes to $x + dx$ and y to $y + dy$ will be fixed, and have no dependence on the law of second differentials. Hence, since $dy/dx = p$, we have $dy = p\,dx$ and $d^2y = pd^2x + dp\,dx$, so that

$$dx\,d^2y - dy\,d^2x = dp\,dx^2,$$

and this value is not unsettled, since it contains only first differentials. This property is consistent whether any differential is taken as constant, whatsoever it might be, or even if no differential is held constant.

260. The formula $dy\,d^2x - dx\,d^2y$ has a fixed signification even though it contains second differentials, which can be thought of as destroying each other. Any expression in which there are no other second differentials besides the formula $dy\,d^2x - dx\,d^2y$ likewise has a fixed meaning. Now, if we let $dy\,d^2x - dx\,d^2y = \omega$ and if V is a quantity formed from x, y, their first differentials dx, dy, and ω, then V will have a fixed value. Since in the first differentials dx and dy there is no reason for uncertainty as to the law by which the successive values change as x increases, and in ω the second differentials cancel each other, so that the quantity V is not uncertain but fixed. Thus the expression

$$\frac{\left(dx^2 + dy^2\right)^{3/2}}{dx\,d^2y - dy\,d^2x}$$

has a fixed value although it seems to be contaminated by second differentials. In addition, since the numerator is homogeneous with the denominator, it has a finite value, unless by chance it becomes infinitely large or infinitely small.

261. Just as the formula $dx\, d^2y - dy\, d^2x$ has a fixed value, as has been shown, so also if a third variable z is added, these formulas $dx\, d^2z - dz\, d^2x$ and $dy\, d^2z - dz\, d^2y$ have fixed values. Hence, expressions in three variables x, y, and z, provided that there are no second differentials except these, then the expression will be fixed, just as if they contained no second differential at all. It follows that this expression

$$\frac{\left(dx^2 + dy^2 + dz^2\right)^{3/2}}{(dx + dz)\, d^2y - (dy + dz)\, d^2x + (dx - dy)\, d^2z},$$

although it does contain second differentials, still it keeps a fixed signification. In a similar way it is possible to exhibit formulas containing many variables in which second differentials do not prevent the formulas from having a fixed significance.

262. Except for formulas of this kind, all others that contain second differentials will give uncertain signification, and for this reason they have no place in calculations. On the other hand, a first differential may be defined to be constant. As soon as any first differential is assumed to be constant, all expressions, no matter how many variables they may contain and no matter what differentials higher than the first may be present, obtain fixed significance, and they are no longer excluded from calculations. For example, if dx is assumed to be constant, the second differential of x and all higher differentials vanish. Whatsoever functions of x may be substituted for the other variables y, z, and so forth, their second differentials through dx^2, their third through dx^3, and so forth, will be determined. In this way the ambiguity is removed from the second differentials. The same thing is true if the first differential of some other variable or function is made constant.

263. From this it follows that second and higher differentials never enter into a calculation, and because of their unsettled signification they are completely unsuitable for analysis. Now, when second differentials seem to be present, either some first differential is assumed to be constant, or this is not the case. In the first case, the second differentials simply vanish, since they are determined by the first differential. In the latter case, unless they cancel each other, the signification is unsettled, and for this reason they have no place in analysis. On the other hand, if they cancel each other, they only seem to be present, and only finite quantities with their first differentials are to be considered really present. However, since they very frequently only seem to be used in calculations, it was necessary that the method of dealing with them be explained. Soon, now, we will show the method by means of which second and higher differentials can always be eradicated.

264. If an expression contains a single variable x and its higher differentials d^2x, d^3x, d^4x, etc. occur in the expression, then it can have no fixed meaning unless some first differential is set constant. Thus, let t be that variable whose differential dt is set constant. Then $d^2t = 0$, $d^3t = 0$, $d^4t = 0$, etc. We let $dx = p\,dt$, and p will be a finite quantity whose differential is not affected by the unsettled signification of second differentials; furthermore, dp/dt will be a finite quantity. Let $dp = q\,dt$, and in a similar way $dq = r\,dt$, $dr = s\,dt$, etc. Here q, r, s, etc. are finite quantities with fixed signification. Since $dx = p\,dt$, we have

$$d^2x = dp\,dt = q\,dt^2,$$

$$d^3x = dq\,dt^2 = r\,dt^3,$$

$$d^4x = dr\,dt^3 = s\,dt^4,$$

$$\ldots\,.$$

If these values are substituted for d^2x, d^3x, d^4x, and so forth, the whole expression will contain only finite expressions and the first differential of dt, nor will there be any unsettled signification.

265. If x were a function of t, then in this way the quantity x could be completely eliminated, so that only the quantity t and its differential dt would remain in the expression. However, if t were a function of x, then x would also be a function of t. Nevertheless, this quantity x with its first differential dx can be retained in the calculation, provided that after the substitutions previously made for t and dt, the values expressed by x and dx are restored. In order that this might become clearer, we will let $t = x^n$, so that the first differential of x^n will be held constant. Since $dt = nx^{n-1}dx$, we have $p = 1/\left(nx^{n-1}\right)$ and

$$dp = \frac{-(n-1)\,dx}{nx^n} = q\,dt = nqx^{n-1}dx,$$

so that

$$q = \frac{-(n-1)}{n^2x^{2n-1}}$$

and

$$dq = \frac{(n-1)(2n-1)\,dx}{n^2x^{2n}} = r\,dt = nrx^{n-1}dx.$$

From this it follows that

$$r = \frac{(n-1)(2n-1)}{n^3x^{3n-1}}$$

and

$$s = \frac{-(n-1)(2n-1)(3n-1)}{n^4 x^{4n-1}}.$$

Hence, if we let the differential of x^n be constant, then

$$d^2x = -\frac{(n-1)\,dx^2}{x},$$

$$d^3x = \frac{(n-1)(2n-1)\,dx^3}{x^2},$$

$$d^4x = -\frac{(n-1)(2n-1)(3n-1)\,dx^4}{x^3},$$

and so forth.

266. If an expression contains two variables x and y, and if the differential of one, x, is held constant, then since $d^2x = 0$, there will be no second or higher differentials besides d^2y, d^3y, etc. However, these can be treated in the same way as before. They can be removed by letting $dy = p\,dx$, $dp = q\,dx$, $dq = r\,dx$, $dr = s\,dx$, and so forth. Then we have

$$d^2y = q\,dx^2, \qquad d^3y = r\,dx^3, \qquad d^4y = s\,dx^4, \qquad \ldots,$$

and so forth. By means of these substitutions we obtain an expression that contains only the differential dx besides the finite quantities x, y, p, q, r, s, etc. For example, if the given expression is

$$\frac{y\,dx^4 + x\,dy\,d^3y + x\,d^4y}{(x^2+y^2)\,d^2y},$$

in which we assume that dx is constant, then we let $dy = p\,dx$, $dp = q\,dx$, $dq = r\,dx$, and $dr = s\,dx$. When these values are substituted the given expression is transformed into

$$\frac{(y + xpr + xs)\,dx^2}{(x^2+y^2)\,q},$$

which contains no second or higher differential.

267. In a similar way the second and higher differentials are removed if dy is assumed to be constant. However, if any other first differential dt is taken to be constant, then the higher differentials of x are removed from the calculation with the method first mentioned before. That is, we let

$$dx = p\,dt, \qquad dp = q\,dt, \qquad dq = r\,dt, \qquad dr = s\,dt, \qquad \ldots,$$

so that

$$d^2x = q\,dt^2, \qquad d^3x = r\,dt^3, \qquad d^4x = s\,dt^4, \qquad \ldots.$$

Then in a similar way for the higher differentials of y we let

$$dy = P\,dt, \qquad dP = Q\,dt, \qquad dQ = R\,dt, \qquad dR = S\,dt, \qquad \ldots,$$

so that

$$d^2y = Q\,dt^2, \qquad d^3y = R\,dt^3, \qquad d^4y = S\,dt^4, \qquad \ldots.$$

When these substitutions are made we obtain an expression that contains only the differential dt besides the finite quantities x, p, q, r, s, etc. y, P, Q, R, S, etc. It follows that there is no unsettled signification.

268. If the first differential that is made constant depends on x or on y, or if it depends on both at the same time, then it is not necessary to introduce a pair of series of finite quantities p, q, r, etc. Indeed, if dt depends only on x, then the letters p, q, r, etc. will be functions of x, and only the letters P, Q, R, etc. will be present. The same thing will occur if the constant differential dt depends only on y. However, if dt depends on both, then the operation must be changed a bit. For example, we let the differential $y\,dx$ be constant, so that $y\,d^2x + dx\,dy = 0$ and

$$d^2x = -\frac{dx\,dy}{y}.$$

Now let

$$dy = p\,dx, \qquad dp = q\,dx, \qquad dq = r\,dx, \qquad \ldots,$$

so that

$$d^2x = -\frac{p\,dx^2}{y}.$$

When we differentiate further, we obtain

$$d^3x = -\frac{q\,dx^3}{y} + \frac{p^2dx^2}{y^2} - \frac{2p\,dx\,d^2x}{y},$$

and when we substitute $-p\,dx^2/y$ for d^2x we have

$$d^3x = -\frac{q\,dx^3}{y} + \frac{3p^2\,dx^2}{y^2}.$$

Furthermore,

$$d^4x = -\frac{r\,dx^4}{y} + \frac{pq\,dx^4}{y^2} + \frac{6pq\,dx^4}{y^2} - \frac{6p^3dx^4}{y^3} + \left(\frac{3p^2}{y^2} - \frac{q}{y}\right) 3dx^2d^2x$$

and when for d^2x we substitute $-p\,dx^2/y$, we have

$$d^4x = \left(\frac{-r}{y} + \frac{10pq}{y^2} - \frac{15p^3}{y^3}\right) dx^4$$

and so forth. Then, since $dy = p\,dx$, we have

$$d^2y = q\,dx^2 + pd^2x = \left(q - \frac{p^2}{y}\right) dx^2,$$

and again substituting $-p\,dx^2/y$ for d^2x we have

$$d^3y = \left(r - \frac{4pq}{y} + \frac{3p^3}{y^2}\right) dx^3$$

and

$$d^4y = \left(s - \frac{7pr}{y} - \frac{4q^2}{y} + \frac{25p^2q}{y^2} - \frac{15p^4}{y^3}\right) dx^4,$$

and so forth. When these values are substituted for the higher differentials of x and y, a given expression is transformed into a form of the kind which no longer contains higher differentials. This is accomplished by considering some differential to be constant.

269. Frequently in the application of calculus to curves it may happen that the first differential $\sqrt{dx^2 + dy^2}$ is assumed to be constant. For this reason we now show the way in which for this case the second and higher differentials should be eliminated. At the same time, by using the same argument, the way will be opened to show the process if any other differential is assumed to be constant. Now we let

$$dy = p\,dx, \qquad dp = q\,dx, \qquad dq = r\,dx, \qquad dr = s\,dx, \qquad \ldots.$$

Then the differential $\sqrt{dx^2 + dy^2}$ takes the form $dx\sqrt{1 + p^2}$. Since this is constant, we have

$$d^2x\sqrt{1 + p^2} + \frac{pq\,dx^2}{\sqrt{1 + p^2}} = 0,$$

so that

$$d^2x = -\frac{pq\,dx^2}{1 + p^2},$$

and we already have the value of d^2x. Furthermore, we have

$$\begin{aligned} d^3x &= -\frac{pr\,dx^3}{1 + p^2} - \frac{q^2dx^3}{1 + p^2} + \frac{2p^2q^2dx^3}{(1 + p^2)^2} - \frac{2pq\,dx\,d^2x}{1 + p^2} \\ &= -\frac{pr\,dx^3}{1 + p^2} - \frac{q^2dx^3}{1 + p^2} + \frac{4p^2q^2dx^3}{(1 + p^2)^2} = -\frac{pr\,dx^3}{1 + p^2} + \frac{(3p^2 - 1)\,q^2dx^3}{(1 + p^2)^2}. \end{aligned}$$

Then we have

$$d^4x = -\frac{ps\,dx^4}{1+p^2} + \frac{\left(10p^2-3\right)qr\,dx^4}{\left(1+p^2\right)^2} - \frac{\left(15p^2-13\right)pq^3dx^4}{\left(1+p^2\right)^3}.$$

Since we are assuming that $dy = p\,dx$, when this is differentiated, we have

$$d^2y = q\,dx^2 + pd^2x = q\,dx^2 - \frac{p^2q\,dx^2}{1+p^2} = \frac{q\,dx^2}{1+p^2},$$

$$d^3y = \frac{r\,dx^3}{1+p^2} - \frac{2pq^2dx^3}{\left(1+p^2\right)^2} + \frac{2q\,dx\,d^2x}{1+p^2},$$

so that

$$d^3y = \frac{r\,dx^3}{1+p^2} - \frac{4pq^2dx^3}{\left(1+p^2\right)^2}.$$

When we differentiate again we have

$$d^4y = \frac{s\,dx^4}{1+p^2} - \frac{13pqr\,dx^4}{\left(1+p^2\right)^2} + \frac{4\left(6p^2-1\right)q^3dx^4}{\left(1+p^2\right)^3}.$$

Hence all higher differentials of both x and y are expressed through finite quantities and powers of dx. After these substitutions, the resulting expression is completely free of second differentials.

270. Now that we have given the method for stripping second and higher differentials from expressions, it is fitting that we illustrate this material with some few examples.

I. Let the given expression be $x\,d^2y/dx^2$, in which dx is set constant. Hence we let $dy = p\,dx$ and $dp = q\,dx$, so that $d^2y = q\,dx^2$ and the given expression becomes this finite quantity xq.

II. Let the given expression be $\left(dx^2+dy^2\right)/d^2x$, in which dy is set constant. We let $dx = pdy$, $dp = q\,dy$. Since $d^2x = q\,dy^2$, we obtain $\left(1+p^2\right)/q$. However, if we should wish, as before, to let $dy = p\,dx$, $dp = q\,dx$, since dy is constant, we have $0 = pd^2x + dp\,dx$ and $d^2x = -q\,dx^2/p$. Hence the given expression becomes $-p\left(1+p^2\right)/q$.

III. Let the given expression be

$$\frac{y\,d^2x - x\,d^2y}{dx\,dy},$$

in which $y\,dx$ is set constant. We let $dy = p\,dx$ and $dp = q\,dx$; from paragraph 268 we have $d^2x = -p\,dx^2/y$ and

$$d^2y = q\,dx^2 - \frac{p^2dx^2}{y}.$$

When these are substituted into the given expression, it is transformed into

$$-1 - \frac{xq}{p} + \frac{xp}{y}.$$

IV. Let the given expression be

$$\frac{dx^2 + dy^2}{d^2y},$$

in which we let $\sqrt{dx^2 + dy^2}$ be constant. Again we let $dy = p\,dx$, $dp = q\,dx$, and from the preceding paragraph we have

$$d^2y = \frac{q\,dx^2}{1+p^2}.$$

Hence the given expression becomes $\left(1+p^2\right)^2/q$.

From these examples it should be sufficiently clear, in any given case, the way in which second and higher differentials should be eliminated when any first differential is assumed to be constant.

271. Since second and higher differentials can be eliminated by introducing finite quantities p, q, r, s, etc., so that the whole expression is made up only of the differential dx and the finite quantities p, q, r, s, etc., if an expression reduced in this manner is given, we can again recover the original form by substituting second and higher differentials for the letters p, q, r, s, etc. Now in the same way, some first differential is assumed to be constant, whether it be the one originally so assumed, or some other. However, it could be that no differential was assumed to be constant while it contains second and higher differentials and at the same time it has a fixed signification. We have seen expressions of this kind above.

272. Now let any given expression contain the finite letters x, y, p, q, r, etc. with one differential dx, in which

$$p = \frac{dy}{dx}, \qquad q = \frac{dp}{dx}, \qquad r = \frac{dq}{dx}, \qquad \ldots .$$

If we wish to eliminate these letters, in their place we introduce second and higher differentials of x and y with no differential assumed to be constant. Since

$$dp = \frac{dx\,d^2y - dy\,d^2x}{dx^2},$$

so that

$$q = \frac{dx\,d^2y - dy\,d^2x}{dx^3},$$

this formula gives

$$dq = \frac{dx^2 d^3y - 3dx\, d^2x\, d^2y + 3dy\, d^2x^2 - dx\, dy\, d^3x}{dx^4},$$

so that

$$r = \frac{dx^2 d^3y - 3dx\, d^2x\, d^2y + 3dy\, d^2x^2 - dx\, dy\, d^3x}{dx^5}.$$

Furthermore, if the letter s, which indicates the value of dr/dx, is in the expression, then

$$s = \frac{dx^3 d^4y - 6dx^2 d^2x d^3y - 4dx^2 d^2y\, d^3x + 15dx\, d^2x^2 d^2y}{dx^7}$$
$$+ \frac{10dx\, dy\, d^2x d^3x - 15dy\, d^2x^3 - dx^2 dy\, d^4x}{dx^7}.$$

When these values are substituted for p, q, r, s, etc., into the given expression, that expression is transformed into another one that contains higher differentials of x and y. Even though no first differential is assumed to be constant, still the expression has a fixed signification.

273. In this way any formula for a higher differential in which some first differential is assumed to be constant can be transformed into another form, in which no differential is set equal to a constant, and in spite of this it still has a fixed value. First, by means of the method already discussed, we take the values $dy = p\,dx$, $dp = q\,dx$, $dq = r\,dx$, $dr = s\,dx$, etc., and the higher differentials are eliminated. Then for p, q, r, s, etc., we substitute the values just discovered and this transformation is illustrated by the following examples.

I. Let the given expression be $x\,d^2y/dx^2$, in which we let dx be constant. We would like to transform this into another form that involves no constant differential. We let $dy = p\,dx$, $dp = q\,dx$, and, as seen before in paragraph 270, the given expression becomes qx. Now for q we substitute the value we obtain when no differential is constant, namely,

$$q = \frac{dx\, d^2y - dy\, d^2x}{dx^3}.$$

The resulting expression is then equal to

$$\frac{x\, dx\, d^2y - x\, dy\, d^2x}{dx^3},$$

and this involves no other constant differential.

II. Let the given expression be

$$\frac{dx^2 + dy^2}{d^2x},$$

in which dy is assumed to be constant. We let $dy = p\,dx$ and $dp = q\,dx$, so that the expression becomes $-p\left(1+p^2\right)/q$, as in paragraph 270. Since

$$p = \frac{dy}{dx} \qquad \text{and} \qquad q = \frac{dx\,d^2y - dy\,d^2x}{dx^3},$$

we obtain the expression

$$\frac{dy\left(dx^2 + dy^2\right)}{dy\,d^2x - dx\,d^2y}.$$

Here no differential is assumed constant, and this expression has the same value as the one originally proposed.

III. Let the given expression be

$$\frac{y\,d^2x - x\,d^2y}{dx\,dy},$$

in which the differential $y\,dx$ is assumed to be constant. We let $dy = p\,dx$, $dp = q\,dx$, and as we saw in paragraph 270, this expression is transformed into

$$-1 - \frac{xq}{p} + \frac{xp}{y}.$$

When we do not assume any differential to be constant, the expression is transformed into

$$-1 - \frac{x\,dx\,d^2y - x\,dy\,d^2x}{dx^2dy} + \frac{x\,dy}{y\,dx}$$
$$= \frac{x\,dx\,dy^2 - y\,dx^2dy - yx\,dx\,d^2y + yx\,dy\,d^2x}{y\,dx^2dy}.$$

IV. Let the given expression be

$$\frac{dx^2 + dy2}{d^2y},$$

in which we assume that the differential $\sqrt{dx^2+dy^2}$ is constant. When we let $dy = p\,dx$ and $dp = q\,dx$, there arises the expression $\left(1+p^2\right)^2/q$ as in paragraph 270. Now we set $p = dy/dx$ and

$$q = \frac{dx\,d^2y - dy\,d^2x}{dx^3}$$

with no differential being constant, and the expression

$$\frac{\left(dx^2+dy^2\right)^2}{dx^2d^2y - dx\,dy\,d^2x}$$

becomes equivalent to the proposed expression.

V. Let the given expression be $dx\,d^3y/d^2y$, in which the differential dx is assumed to be constant. We let $dy = p\,dx$, $dp = q\,dx$, and $dq = r\,dx$. Since $d^2y = q\,dx^2$ and $d^3y = r\,dx^3$, the given formula becomes $r\,dx^2/q$. Now for q and r we substitute those values that they receive when no differential is assumed to be constant, that is,

$$q = \frac{dx\,d^2y - dy\,d^2x}{dx^3}$$

and

$$r = \frac{dx^2d^3y - 3dx\,d^2x\,d^2y + 3dy\,d^2x^2 - dx\,dy\,d^3x}{dx^5}.$$

We then obtain the following expression, which is equivalent to that originally given:

$$\frac{dx^2d^3y - 3dx\,d^2x\,d^2y + 3dy\,d^2x^2 - dx\,dy\,d^3x}{dx\,d^2y - dy\,d^2x}$$

$$= \frac{dx\left(dx\,d^3y - dy\,d^3x\right)}{dx\,d^2y - dy\,d^2x} - 3d^2x.$$

274. If we consider these transformations more carefully, we can find a more expeditious method in which it is not necessary to resort to the letters p, q, r, etc. Depending on which differential in the formula is assumed to be constant, different methods are used. First, suppose that the constant differential is dx. When we have substituted $p\,dx$ for dy and conversely dy/dx for p, whenever the differentials dx or dy occur, they are retained without alteration. However, wherever d^2y occurs, after we have substituted $q\,dx^2$ and then for q we have written the value

$$\frac{dx\,d^2y - dy\,d^2x}{dx} \quad \text{or} \quad d^2y - \frac{dy\,d^2x}{dx},$$

the transformation is complete. Furthermore, if in the given expression d^3y occurs, since we have substituted $r\,dx^3$, because of the value already found for r, whenever d^3y is found we write

$$d^3y - \frac{3d^2x\,d^2y}{dx} + \frac{3dy\,d^2x^2}{dx^2} - \frac{dy\,d^3x}{dx}.$$

When this is done, the given expression is transformed into a different one that involves no constant differential. For example, if the given expression is

$$\frac{\left(dx^2 + dy^2\right)^{3/2}}{dx\, d^2y}$$

and dx is set constant, when

$$d^2y - \frac{dy\, d^2x}{dx}$$

is substituted for d^2y, the new form with no constant differential is

$$\frac{\left(dx^2 + dy^2\right)^{3/2}}{dx\, d^2y - dy\, d^2x}.$$

275. From this it is easily gathered that whenever in some expression the differential dy is constant, then wherever we find d^2x we should write

$$d^2x - \frac{dx\, d^2y}{dy}$$

and for d^3x we write

$$d^3x - \frac{3d^2x\, d^2y}{dy} + \frac{3dx\, d^2y^2}{dy^2} - \frac{dx\, d^3y}{dy},$$

in order to obtain an equivalent expression in which no differential is set constant. However, if in the given expression $y\, dx$ is assumed constant, then according to paragraph 268, we have

$$d^2x = -\frac{p\, dx^2}{y} \qquad \text{and} \qquad d^2y = q\, dx^2 - \frac{p^2 dx^2}{y}.$$

In place of d^2x we should everywhere write $-dx\, dy/y$ and in place of d^2y we should everywhere write

$$d^2y - \frac{dy\, d^2x}{dx} - \frac{dy^2}{y}.$$

Since the higher differentials seldom occur in this business, we will progress no further. However, if in the given expression the differential $\sqrt{dx^2 + dy^2}$ is assumed constant, since in paragraph 269 we obtained

$$d^2x = -\frac{pq\, dx^2}{1+p^2} \qquad \text{and} \qquad d^2y = \frac{q\, dx^2}{1+p^2},$$

for d^2x we should everywhere write

$$\frac{dy^2 d^2x - dx\, dy\, d^2y}{dx^2 + dy^2},$$

and for d^2y we everywhere write

$$\frac{dx^2 d^2y - dx\,dy\,d^2x}{dx^2 + dy^2}.$$

Hence, if the given expression is

$$\frac{dy\sqrt{dx^2 + dy^2}}{d^2x},$$

in which $\sqrt{dx^2 + dy^2}$ is assumed to be constant, then the expression is transformed into

$$\frac{\left(dx^2 + dy^2\right)^{3/2}}{dy\,d^2x - dx\,d^2y},$$

in which no differential is assumed to be constant.

276. In order that these reductions can be used more easily, we have brought them together in the following table.

A differential formula of higher order can be transformed into one that involves no constant differential by means of the following substitutions:

I. If the differential dx is assumed to be constant, then for d^2y we write

$$d^2y - \frac{dy\,d^2x}{dx}$$

and for d^3y we write

$$d^3y - \frac{3d^2x\,d^2y}{dx} + \frac{3dy\,d^2x^2}{dx^2} - \frac{dy\,d^3x}{dx}.$$

II. If the differential dy is assumed to be constant, then for d^2x we write

$$d^2x - \frac{dx\,d^2y}{dy},$$

and for d^3x we write

$$d^3x - \frac{3d^2x\,d^2y}{dy} + \frac{3dx\,d^2y^2}{dy^2} - \frac{dx\,d^3y}{dy}.$$

III. If the differential $y\,dx$ is assumed to be constant, then for d^2x we write

$$-\frac{dx\,dy}{y},$$

for d^2y we write

$$d^2y - \frac{dy\,d^2x}{dx} - \frac{dy^2}{y},$$

for d^3x we write

$$\frac{dy\,d^2x}{y} - \frac{dx\,d^2y}{y} + \frac{3dx\,dy^2}{y^2},$$

for d^3y we write

$$d^3y - \frac{3d^2x\,d^2y}{dx} + \frac{3dy\,d^2x^2}{dx^2} - \frac{dy\,d^3x}{dx} - \frac{4dy\,d^2y}{y} + \frac{4dy^2d^2x}{y\,dx} + \frac{3dy^3}{y^2}.$$

IV. If the differential $\sqrt{dx^2+dy^2}$ is assumed to be constant, then for d^2x we write

$$\frac{dy^2d^2x - dx\,dy\,d^2y}{dx^2+dy^2},$$

for d^2y we write

$$\frac{dx^2d^2y - dx\,dy\,d^2x}{dx^2+dy^2},$$

for d^3x we write

$$\frac{dy^2d^3x - dx\,dy\,d^3y}{dx^2+dy^2} + \frac{\left(dx\,d^2y - dy\,d^2x\right)\left(3dy^2d^2y - dx^2d^2y + 4dx\,dy\,d^2x\right)}{\left(dx^2+dy^2\right)^2},$$

for d^3y we write

$$\frac{dx^2d^3y - dx\,dy\,d^3x}{dx^2+dy^2} + \frac{\left(dy\,d^2x - dx\,d^2y\right)\left(3dx^2d^2x - dy^2d^2x + 4dx\,dy\,d^2y\right)}{\left(dx^2+dy^2\right)^2}.$$

277. These expressions, which include no constant differential, are given in such a way that one has freedom to choose any differential to be constant. Hence, differential expressions of higher order in which no differential is assumed to be constant, can be tested to decide whether they have unsettled or fixed significance. We choose arbitrarily some differential, for example dx, to be constant. Then, with the rule provided in the previous paragraph, the expression is reduced once more to a form in which no differential is assumed to be constant. If this agrees with the given expression, then it has fixed significance and does not depend on the variability of second differentials. However, if we obtain a different expression, then the given expression has unsettled significance. Thus, if the given expression is $y\,d^2x - x\,d^2y$, in which no differential is set constant, we investigate whether the significance is unsettled or fixed. We let dx be constant, so that the expression becomes $-x\,d^2y$. Now, with the first rule of the previous paragraph, we substitute

$$d^2y - \frac{dy\,d^2x}{dx}$$

for d^2y and obtain

$$-x\,d^2y + \frac{x\,dy\,d^2x}{dx}.$$

Since these two expressions do not agree, this indicates that the given expression has no fixed and stated significance.

278. In a similar way if the given general expression is of the type

$$P\,d^2x + Q\,dx\,dy + R\,d^2y,$$

it is possible to define a condition under which the expression will have a fixed value even though no differential is assumed to be constant. When dx is set constant, the expression becomes $Q\,dx\,dy + R\,d^2y$. Now this is once more transformed into another form, so that its signification remains the same, even though no differential is thought to be constant. In this way we obtain

$$Q\,dx\,dy + R\,d^2y - \frac{R\,dy\,d^2x}{dx},$$

which agrees with the given expression, provided that $P\,dx + R\,dy = 0$. Only in this case will the value of the expression be fixed. Indeed, if P is not equal to $-R\,dy/dx$ or if R is not equal to $-P\,dx/dy$, the given expression $P\,d^2x + Q\,dx\,dy + R\,d^2y$ has no fixed value. Its signification will be unsettled and vary depending on which differential is assumed to be constant.

279. Using these principles it will be easy to convert a differential expression in which some differential is set constant into another form in which a different differential is assumed to be constant. We reduce the first expression to the form that involves no constant differential. Once this is accomplished, we set the other differential constant. Thus if in the proposed expression the differential dx is assumed to be constant, and this is transformed into another that involves a constant dy, in the above formulas instead of d^2y and d^3y, since dy is constant, we would let $d^2y = 0$, $d^3y = 0$, but the desired result is obtained if for d^2y we substitute

$$\frac{-dy\,d^2x}{dx} \quad \text{and} \quad \frac{3dy\,d^2x}{dx^2} - \frac{dy\,d^3x}{dx}$$

for d^3y. In this way the formula

$$\frac{-\left(dx^2+dy^2\right)^{3/2}}{dx\,d^2y},$$

in which dx is set constant, is transformed into

$$\frac{\left(dx^2+dy^2\right)^{3/2}}{dy\,d^2x},$$

in which dy is set constant.

280. If, on the other hand, a formula in which dy is set constant is to be transformed into another in which dx is constant, then for d^2x we have to substitute

$$\frac{-dx\,d^2y}{dy},$$

and for d^3x the expression

$$\frac{3dx\,d^2y^2}{dy^2}-\frac{dx\,d^3y}{dy}.$$

In a similar way, if a formula in which $\sqrt{dx^2+dy^2}$ is set constant is to be transformed into another in which dx is constant, then for d^2x we write

$$\frac{-dx\,dy\,d^2y}{dx^2+dy^2},$$

and for d^2y we write

$$\frac{dx^2d^2y}{dx^2+dy^2}.$$

However, if a formula in which dx is assumed to be constant is to be transformed into another in which $\sqrt{dx^2+dy^2}$ is to be constant, since dx^2+dy^2 is constant, we have $dx\,d^2x+dy\,d^2y=0$ and

$$d^2x=-\frac{dy\,d^2y}{dx}.$$

This value is given to d^2x, and for d^2y we write

$$d^2y+\frac{dy^2d^2y}{dx^2}=\frac{\left(dx^2+dy^2\right)d^2y}{dx^2}.$$

Hence the formula

$$\frac{-\left(dx^2+dy^2\right)^{3/2}}{dx\,d^2y},$$

in which dx is constant, is transformed into another in which $\sqrt{dx^2+dy^2}$ is set constant, which is

$$\frac{-dx\sqrt{dx^2+dy^2}}{d^2y}.$$

9

On Differential Equations

281. In this chapter we principally set forth an explanation of the differentiation of those functions of x that are not defined explicitly, but implicitly by means of the relationship of x to the function y. Once this is accomplished, we consider the nature of differential equations in general, and we show how they arise from finite equations. Since the main concern in integral calculus is the solution of differential equations, that is finding finite equations that correspond to the differentials, it is necessary here to examine very carefully the nature and properties of differential equations that follow from their origin. In this way we will be preparing the way for integral calculus.

282. In order that we complete this task, let y be a function of x that is defined by this quadratic expression:

$$y^2 + Py + Q = 0.$$

Since this expression y^2+Py+Q is equal to zero, whatever x might signify, the equation will still be equal to zero if we substitute $x+dx$ for x. In this case y becomes $y + dy$. When this substitution is made and the original $y^2 + Py + Q$ is subtracted from the new quantity, there remains the differential, which is also equal to zero. From this it should be clear that if any expression is equal to zero, then its differential will also be equal to zero. Furthermore, if two expressions are equal to each other, then their differentials will be equal. Since $y^2 + Py + Q = 0$, we also have

$$2y\,dy + P\,dy + y\,dP + dQ = 0.$$

Since P and Q are functions of x, their differentials have the form $dP = p\,dx$ and $dQ = q\,dx$ so that

$$2y\,dy + P\,dy + yp\,dx + q\,dx = 0,$$

with the result that

$$\frac{dy}{dx} = -\frac{yp+q}{2y+P}.$$

283. Now, just as the finite equation $y^2+Py+Q=0$ gives the relationship between y and x, so the differential equation expresses the relationship, or the ratio of dy to dx. However since

$$\frac{dy}{dx} = \frac{-yp+q}{2y+P},$$

we cannot know this ratio $dy : dx$ unless we know the function y. Things could not possibly be otherwise, since from the finite equation y has two values. Either of these two values has its own proper differential, and either differential will appear depending on which value is substituted for y. In a similar way, the function y can be defined by a cubic equation. In this case, dy/dx will have three values, depending on which of the three values of y is substituted. If in a given finite equation y has four or more values, then of necessity, dy/dx has just as many significations.

284. Nevertheless, the function y can be eliminated, since we have two equations containing y, namely the finite and the differential. In that case, however, the differential dy will take on as many values as y had in the original finite equation. Thus there are this many ratios of dy to dx. Let us take the preceding example, $y^2 + Py + Q = 0$, whose differential is $2y\,dy + P\,dy + y\,dP + dQ = 0$, from which we obtain

$$y = -\frac{P\,dy + dQ}{2dy + dP}.$$

When this value for y is substituted in the previous equation, we have

$$\left(4Q - P^2\right) dy^2 + \left(4Q - P^2\right) dP\,dy + Q\,dP^2 - P\,dP\,dQ + dQ^2 = 0,$$

whose roots are

$$dy = -\frac{1}{2}dP \pm \frac{\frac{1}{2}P\,dP - dQ}{\sqrt{P^2 - 4Q}}.$$

These two differentials correspond to the two values of y from the original finite equation

$$y = -\frac{1}{2}P + \frac{1}{2}\sqrt{P^2 - 4Q}.$$

285. Once the value of dy has been found, by repeated differentiation we find the value of d^2y, d^3y, d^4y, etc. Since these have no determined value unless some first differential is made constant, for convenience, let dx be constant, and for an illustration, let us consider this example:

$$y^3 + x^3 = 3axy.$$

By differentiation we obtain

$$3y^2dy + 3x^2dx = 3ax\,dy + 3ay\,dx,$$

so that

$$\frac{dy}{dx} = \frac{ay - x^2}{y^2 - ax}.$$

Once more we take differentials with dx constant and obtain

$$\frac{d^2y}{dx} = \frac{-ay^2dy - a^2x\,dy + 2x^2y\,dy - 2xy^2dx + a^2y\,dx + ax^2dx}{\left(y^2 - ax\right)^2}.$$

When we substitute for dy the value already obtained,

$$\frac{ay\,dx - x^2dx}{y^2 - ax},$$

and divide by dx, we have

$$\frac{d^2y}{dx^2} = \frac{\left(ay - x^2\right)\left(2x^2y - ay^2 - a^2x\right)}{\left(y^2 - ax\right)^3}$$

$$+ \frac{ax^2 + a^2y - 2xy^2}{\left(y^2 - ax\right)^2},$$

or

$$\frac{d^2y}{dx^2} = \frac{6ax^2y^2 - 2x^4y - 2xy^4 - 2a^3xy}{\left(y^2 - ax\right)^3}$$

$$= -\frac{2a^3xy}{\left(y^2 - ax\right)^3},$$

since from the finite equation we have $2x^4y + 2xy^4 = 6ax^2y^2$. In this way, using the finite equation, these values can be transformed into innumerable different forms.

286. A differential equation can be expressed in an infinite number of ways, since it can be combined with the finite equation. Thus, with the preceding example we obtained the differential equation

$$y^2dy + x^2dx = ax\,dy + ay\,dx,$$

and if this is multiplied by y, we have

$$y^3 dy + x^2 y\,dx = axy\,dy + ay^2 dx.$$

If we substitute for y^3 its value $3axy - x^3$, we obtain the new equation

$$2ax\,dy - x^3 dy + x^2 y\,dx = ay^2 dx.$$

If we multiply this equation again by y and then substitute for y^3 its value, we have

$$2axy^2 dy - x^3 y\,dy + x^2 y^2 dx = 3a^2 xy\,dx - ax^3 dx.$$

In general, if P, Q, R represent any functions of x and y, and if the differential equation is multiplied by P, then

$$Py^2 dy + Px^2 dx = aPx\,dy + aPy\,dx.$$

Now, since $x^3 + y^3 - 3axy = 0$, we also have

$$\left(x^3 + y^3 - 3axy\right)(Q\,dx + R\,dy) = 0.$$

When these equations are added to each other we obtain a general differential equation that arises from the given finite equation

$$Py^2 dy - aPx\,dy + Rx^3 dy + Ry^3 dy - 3aRxy\,dy$$
$$+ Px^2 dx - aPy\,dx + Qx^3 dx + Qy^3 dx - 3aQxy\,dx = 0.$$

287. It is possible to find an infinite number of differential equations through differentiation from the same finite equation, since before differentiation the equation can be multiplied or divided by an arbitrary quantity. Thus, if P is any function of x and y, so that $dP = p\,dx + q\,dy$, and if the finite equation is multiplied by P and then differentiated, we obtain a general differential equation that takes on an infinite number of forms insofar as P takes on one or another function. This multiplicity is increased infinitely if this equation is added to the original finite equation multiplied by the formula $Q\,dx + R\,dy$, where Q and R can be any functions of x and y. Although in all of these equations the relation between dy and dx remains, and this is determined by the differential of the function y in the original finite equation, nevertheless, the differential of y could be determined by other finite equations. The reason for this is better explained in integral calculus.

288. Not only can we obtain an innumerable number of equations from a single finite equation, but we can also find an infinite number of finite equations that lead to the same differential equation. For instance, these two equations $y^2 = ax + ab$ and $y^2 = ax$ are entirely different, since in the first any constant quantity can be given to b. Nevertheless, when both equations are differentiated, we obtain the same differential equation

$$2y\,dy = a\,dx.$$

Indeed, all of the equations represented by $y^2 = ax$, depending on the value assigned to a, correspond to a single differential equation that contains no a. If this equation is divided by x, so that $y^2/x = a$, then this, when differentiated, gives

$$2x\,dy - y\,dx = 0.$$

Even both transcendental and algebraic equations can lead to the same differential equation, as is seen in the equations

$$y^2 - ax = 0 \qquad \text{and} \qquad y^2 - ax = b^2 e^{x/a}.$$

If each of these equations is divided by $e^{x/a}$, so that we have

$$e^{-x/a}\left(y^2 - ax\right) = 0 \qquad \text{and} \qquad e^{-x/a}\left(y^2 - ax\right) = b^2,$$

then when each of these is differentiated, the same differential equation results:

$$2y\,dy - a\,dx - \frac{y^2dx}{a} + x\,dx = 0.$$

289. The reason for this diversity is the fact that the differential of a constant quantity is equal to zero. Hence, if a finite equation can be reduced to such a form that some constant quantity stands alone, neither multiplied nor divided by variables, then by differentiation there arises an equation in which that constant quantity is completely eliminated. In this way any constant quantity involved in a finite equation can be eliminated through differentiation. Thus, if the given equation is

$$x^3 + y^3 = 3axy,$$

and if this is divided by xy, so that we have

$$\frac{x^3 + y^3}{xy} = 3a,$$

then when this equation is differentiated we have

$$2x^3y\,dx + 2xy^3dy - x^4dy - y^4dx = 0,$$

in which the constant no longer appears.

290. If we need to remove several constant quantities from some finite equation, we accomplish this through differentiating two or more times, and in this way we finally obtain differential equations of higher orders in which the constants have been completely removed. Let the given equation be

$$y^2 = ma^2 - nx^2,$$

from which we need to remove the constants ma^2 and n. We remove the first by differentiation, to obtain

$$y\,dy + nx\,dx = 0.$$

From this we form the equation

$$\frac{y\,dy}{x\,dx} + n = 0.$$

When we take dx to be constant, through differentiation, we have

$$xy\,d^2y + x\,dy^2 - y\,dx\,dy = 0.$$

This equation, although it contains no constant, still results from every equation that has the form $y^2 = ma^2 - nx$, no matter what values may be assigned to the letters m, n, and a^2.

291. Not only constant quantities can be eliminated by differentiation from finite equations, but also some variables, namely that variable whose differential is assumed to be constant can be eliminated by differentiation. Indeed, from a given equation in x and y, let us find the value of x such that $x = Y$, where Y is a function of y. Then $dx = dY$, and when dx is taken to be constant, by differentiation we have $0 = d^2Y$. However, if

$$x^2 + ax + b = Y,$$

then by three differentiations we have $0 = d^3Y$. By differentiation four times the equation

$$x^3 + ax^2 + bx + c = Y$$

gives $0 = d^4Y$. In all of these equations, although only one variable seems to be present, while another variable can be missing from the equation, still, since the differential dx is assumed to be constant, we must in reality remember that there is some relationship to x and consider x as belonging to the equation. Hence it should cause no surprise if frequently differential equations of second or higher order occur in which only one variable seems to be involved.

292. It is particularly important to notice that irrational and transcendental quantities can be eliminated from an equation by differentiation. With regard to irrationals, by known methods of reduction irrationals can be eliminated, and once this is accomplished, by differentiation we obtain an equation free of any irrationality. However, frequently it can be more convenient without the reduction to remove the irrationality by comparing the differential equation with the finite formula, provided that there is only one irrational quantity. If there are two or more irrational parts in the finite equation, then the differential equation is differentiated again as many times as there are individual irrational parts to be eliminated, and hence the differential equation will be of a higher order. In this way arbitrary exponents and fractional exponents can also be eliminated. For example, if

$$y^m = \left(a^2 - x^2\right)^n,$$

then after differentiation we have

$$my^{m-1}dy = -2n\left(a^2 - x^2\right)^{n-1} x\,dx.$$

When this equation is divided by the finite equation we have

$$\frac{m\,dy}{y} = -\frac{2nx\,dx}{a^2 - x^2},$$

in which there remains no arbitrary exponent. It should be clear that a differential equation that is free of all irrationality can arise from a finite equation that involves an irrationality or even a transcendental quantity.

293. In order that we understand the way in which transcendental quantities are eliminated by differentiation we begin with logarithms. Since the differential of a logarithm is algebraic, this operation causes no difficulty. Thus, if

$$y = x \ln x,$$

then $y/x = \ln x$, and by differentiation we have

$$\frac{x\,dy - y\,dx}{x^2} = \frac{dx}{x},$$

so that

$$x\,dy - y\,dx = x\,dx.$$

If there are two logarithms, then two differentiations are required. If

$$y \ln x = x \ln y,$$

then

$$\frac{y \ln x}{x} = \ln y,$$

and by differentiation,

$$\frac{x\,dy\ln x + y\,dx - y\,dx\ln x}{x^2} = \frac{dy}{y}.$$

We conclude that

$$\ln x = \frac{x^2dy - y^2dx}{yx\,dy - y^2dx}.$$

When this equation is differentiated, with dx set constant, we have

$$\frac{dx}{x} = \frac{x^2d^2y + 2x\,dx\,dy - 2y\,dx\,dy}{yx\,dy - y^2dx} + \frac{\left(y^2dx - x^2dy\right)\left(yx\,d^2y + x\,dy^2 - y\,dx\,dy\right)}{\left(yx\,dy - y^2dx\right)^2},$$

or

$$\frac{dx}{x} = \frac{y^3x\,dx\,d^2y - y^2x^2dx\,d^2y + 3yx^2dx\,dy^2}{\left(yx\,dy - y^2dx\right)^2} + \frac{-y^2x\,dx\,dy^2 + y^3dx^2dy - 2xy^2dx^2dy - x^3dy^3}{\left(yx\,dy - y^2dx\right)^2},$$

which by reduction gives

$$y^3x\,dx\,d^2y - y^2x^2dx\,d^2y + 3yx^2dx\,dy^2 - 2xy^2dx\,dy^2 + 3y^3dx^2dy - 2xy^2dx^2dy - x^3dy^3 - \frac{y^4dx^3}{x} = 0,$$

or

$$y^2x^2\left(y - x\right)dx\,d^2y + 3yx\,dx\,dy\left(x^2dy + y^2dx\right) - 2y^2x^2dx\,dy\left(dx + dy\right) = x^4dy^3 + y^4dx^3.$$

294. Exponential quantities are removed by differentiation in the same way as logarithms. If the given equation is

$$P = e^Q,$$

where P and Q are any functions of x and y, the equation can be transformed into the logarithmic equation $\ln P = Q$, whose differential is

$$\frac{dP}{P} = Q, \qquad \text{or} \qquad dP = P\,dQ.$$

There is no real difficulty if the exponential quantities are more complicated. In this case, if one differentiation is not sufficient, then two or more differentiations will solve the problem.

I. Let

$$y = \frac{e^x + e^{-x}}{e^x - e^{-x}}.$$

When both numerator and denominator are multiplied by e^x, we have

$$y = \frac{e^{2x} + 1}{e^{2x} - 1},$$

so that

$$e^{2x} = \frac{y+1}{y-1} \quad \text{and} \quad 2x = \ln\left(\frac{y+1}{y-1}\right),$$

whose differential is

$$dx = -\frac{dy}{y^2 - 1} = \frac{dy}{1 - y^2}.$$

II. Let

$$y = \ln\left(\frac{e^x + e^{-x}}{2}\right).$$

By the first differentiation

$$dy = \frac{e^x - e^{-x}}{e^x + e^{-x}} dx,$$

or

$$\frac{dy}{dx} = \frac{e^{2x} - 1}{e^{2x} + 1} \quad \text{and} \quad e^{2x} = \frac{dy + dx}{dx - dy}.$$

Hence

$$2x = \ln\left(\frac{dy + dx}{dx - dy}\right).$$

If we take dx to be constant, then

$$dx = \frac{dx\, d^2y}{dx^2 - dy^2},$$

or

$$dx^2 = d^2y + dy^2.$$

295. In a similar way trigonometric quantities can be removed from an equation by differentiation, as can be understood from the following examples.

I. Let

$$y = a \arcsin \frac{x}{a}.$$

Then

$$dy = \frac{a\, dx}{\sqrt{a^2 - x^2}}.$$

II. Let

$$y = a \cos \frac{y}{x}.$$

Then

$$\frac{y}{a} = \cos \frac{y}{x} \qquad \text{and} \qquad \frac{dy}{a} = \frac{-x\,dy + y\,dx}{x^2} \sin \frac{y}{x}.$$

Since

$$\cos \frac{y}{x} = \frac{y}{a},$$

we have

$$\sin \frac{y}{x} = \frac{\sqrt{a^2 - y^2}}{a}.$$

When we substitute this value into the differential equation, we have

$$\frac{dy}{a} = \frac{(y\,dx - x\,dy)\sqrt{a^2 - y^2}}{ax^2},$$

or

$$x^2 dy = (y\,dx - x\,dy)\sqrt{a^2 - y^2}.$$

III. Let $y = m \sin x + n \cos x$. After the first differentiation we have $dy = m\,dx \cos x - n\,dx \sin x$. When we keep dx constant and differentiate again the result is $d^2y = -m\,dx^2 \sin x - n\,dx^2 \cos x$. When this equation is divided by the given one we have $d^2y/y = -dx^2$, or

$$d^2y + y\,dx^2 = 0,$$

from which not only the sine and cosine have been eliminated, but also the constants m and n.

IV. Let $y = \sin \ln x$. Then $\arcsin y = \ln x$, and by differentiation we have

$$\frac{dy}{\sqrt{1 - y^2}} = \frac{dx}{x}.$$

When each side is squared, the result is $x^2 dy^2 = dx^2 - y^2 dx^2$. When we let dx be constant, by another differentiation we obtain $2x^2 dy\,d^2y + 2x\,dx\,dy^2 = -2y\,dx^2 dy$, or

$$x^2 d^2y + x\,dx\,dy + y\,dx^2 = 0.$$

V. Let $y = ae^{mx} \sin nx$. Then by differentiation,

$$dy = mae^{mx} dx \sin nx + nae^{mx} dx \cos nx.$$

When this is divided by the given equation we have

$$\frac{dy}{y} = m\,dx + \frac{n\,dx\cos nx}{\sin nx} = m\,dx + n\,dx\cot nx,$$

so that

$$\operatorname{arccot}\left(\frac{dy}{ny\,dx} - \frac{m}{n}\right) = nx.$$

If we let dx be constant and differentiate, then [1]

$$n\,dx = \frac{n\,dx\,dy^2 - ny\,dx\,d^2y}{m^2y^2dx^2 + n^2y^2dx^2 - 2my\,dx\,dy + dy^2},$$

or

$$\left(m^2 + n^2\right)y^2dx^2 - 2my\,dx\,dy = -y\,d^2y.$$

It should be clear that although the differential equation may contain no transcendental quantity, still the finite equation from which it originated may contain transcendental quantities of various kinds.

296. Therefore, differential equations, whether of the first or higher order, which contain two variables, x and y, arise from finite equations that also express a relationship between the two variables. Indeed, given any differential equation containing these two variables x and y, there is expressed a relationship between x and y such that y becomes a function of x. From this we can see the nature of a differential equation. That is, if we can assign to y a function of x that is indicated by the equation and is such that when the function is substituted for y its differential is substituted for dy, and its higher differentials for d^2y, d^3y, etc., then the resulting equation is an identity. Integral calculus is concerned with the investigation of such functions. It has this purpose, that given any differential equation, a function of x should be defined that is equal to the other variable y, or what is equivalent, that a finite equation be found that contains the relationship between x and y.

297. If, for example, the given equation is

$$2y\,dy - a\,dx - \frac{y^2dx}{a} + x\,dx = 0,$$

---

[1] This is a correction by Gerhard Kowalewski. The original edition had

$$n\,dx = \frac{n\,dx\,dy^2 - ny\,dx\,d^2y}{m^2y^2dx^2 + n^2y^2dx^2 - 2my\,dx\,dy}.$$

which we arrived at above in paragraph 288, the same relationship between x and y is defined as that contained in the finite equation

$$y^2 - ax = b^2 e^{x/a}.$$

From this we have $y^2 = ax + b^2 e^{x/a}$, so that

$$\sqrt{ax + b^2 e^{x/a}} = y,$$

which is the function of x that y equals in the given differential equation. Indeed, if we substitute the value $ax + b^2 e^{x/a}$ for y^2 and if we substitute its differential

$$a\,dx + \frac{b^2}{a} e^{x/a} dx$$

for $2y\,dy$, we obtain the following identity:

$$a\,dx + \frac{b^2}{a} e^{x/a} dx - a\,dx - x\,dx - \frac{b^2}{a} e^{x/a} dx + x\,dx = 0.$$

Hence it is clear that every differential equation exhibits the same relationship between x and y as a certain finite relationship, which we can find only with the aid of integral calculus.

298. In order that this may be understood more clearly, we suppose that we know the function of x that is equal to y by reason of the differential equation, whether of the first order or of higher order. We also let

$$dy = p\,dx, \qquad dp = q\,dx, \qquad dq = r\,dx,$$

etc., and if in the equation we take dx to be constant, then $d^2y = q\,dx^2$, $d^3y = r\,dx^3$, etc. When these values are substituted into the differential equation, due to the homogeneous terms, there remains an equation containing only finite quantities x, y, p, q, r, etc. Since p, q, r, etc. are quantities that naturally depend on the function y, there really remains an equation between two variables x and y. In turn it should be clear that a certain relationship between the variables x and y is determined by every differential equation. For this reason, if in the solution of some problem we obtain a differential equation between x and y, we can consider this to be equivalent to a relationship between x and y, just as if we had come to a finite equation.

299. In this way any differential equation can be reduced to finite form, so that it contains nothing but finite quantities when the differentials, that is, the infinitely small quantities, are removed. Since y is some certain function of x, if we let $dy = p\,dx$, $dp = q\,dx$, $dq = r\,dx$, etc., when some differential is taken to be constant, the second and higher differentials are expressed through powers of dx and then completely removed through division. For example, if the given equation is

$$xy\,d^3y + x^2dy\,d^2y + y^2dx\,d^2y - xy\,dx^3 = 0,$$

in which dx is taken as constant, we let $dy = p\,dx$, $dp = q\,dx$, and $dq = r\,dx$, so that the equation becomes

$$xyr + x^2pq + y^2q - xy = 0.$$

After the whole equation is divided by dx^3, this finite equation determines the relationship between x and y.

300. Every differential equation, no matter of what order, by means of the substitutions

$$dy = p\,dx, \qquad dp = q\,dx, \qquad dq = r\,dx,$$

etc., can be reduced to finite quantities. Indeed, if the differential equation is of the first order, so that it contains only the first differential, by means of this reduction, besides y and x, the quantity p is also introduced. If the differential equation is of the second order, containing a second differential, then also the quantity q is introduced; if it be of the third order, then we also have r; and so forth. Since the differentials are eliminated from the calculation in this way, the question about a constant differential has not gone away. Even though we have the quantities q and r arising from second differentials we still have to indicate whether some differential is taken to be constant. It comes to this, whether or not in the development some differential has been arbitrarily taken to be constant.

301. If some differential equation of second or higher order is given, and no constant first differential is indicated, we can explore in the following way whether or not there is a determined relationship between the variables x and y. Since no differential is assumed to be constant, we are free to choose whatever differential we want to be constant. By choosing different differentials to be constant we see whether the same relationship between x and y is given. If this does not happen, then it is a certain sign that the equation expresses no determined relationship, and therefore can have no place in the solution of a problem. However, the safest method, and also the easiest, to explore this question is that given above in paragraph 277. There, in a similar question, we gave a test for determining whether or not differential expressions of higher order have a fixed signification.

302. Hence, given a differential equation of second or higher order, with no differential set constant, we let dx be constant. Then, as we have shown above in paragraph 276 for differential expressions, this equation will be reduced to the same form, which supposes that no differential is constant.

That is, we substitute

$$d^2y - \frac{dy\,d^2x}{dx}$$

for d^2y, and

$$d^3y - \frac{3d^2x\,d^2y}{dx} + \frac{3dy\,d^2x^2}{dx^2} - \frac{dy\,d^3x}{dx}$$

for d^3y, etc. When this is done it becomes clear whether the resulting equation is the same as the given equation. If this is the case, the given equation gives a determined relationship between x and y, just as we have shown at length.

303. In order that this become perfectly clear let us take a given equation in which no constant differential seems to be given:

$$P\,d^2x + Q\,d^2y + R\,dx^2 + S\,dx\,dy + T\,dy^2 = 0.$$

When we let dx be constant, the equation becomes

$$Q\,d^2y + R\,dx^2 + S\,dx\,dy + T\,d^2 = 0.$$

From this now, the consideration of a constant differential is removed in the previously prescribed manner, to obtain

$$-\frac{Q\,dy\,d^2x}{dx} + Q\,d^2y + R\,dx^2 + S\,dx\,dy + T\,dy^2 = 0.$$

Since this equation differs from the original only in the first term, we must see whether $P = -Q\,dy/dx$. If this is the case, we conclude that the given equation exhibits a fixed relationship between x and y, which can be found by the rules of integral calculus, whichever differential is taken to be constant. However, if it is not true that $P = -Q\,dy/dx$, then the given equation is impossible.

304. It follows that unless the given equation

$$P\,d^2x + Q\,d^2y + R\,dx^2 + S\,dx\,dy + T\,dy^2 = 0$$

is meaningless, it is necessary that $P\,dx + Q\,dy = 0$. This can happen in two ways. First, the equation

$$P = -\frac{Q\,dy}{dx}, \qquad \text{or} \qquad P\,dx + Q\,dy = 0,$$

is an identical equation. Second, the equation $P\,dx + Q\,dy = 0$ is itself a first-order differential equation whose differentiation gave rise to the given equation. In the second case, the equation $P\,dx + Q\,dy = 0$ corresponds to the given equation and contains the same relationship between x and y. For this reason the solution can be found without the aid of integral calculus. Indeed, when $P\,dx + Q\,dy = 0$ is differentiated, we obtain

$$P\,d^2x + Q\,d^2y + dP\,dx + dQ\,dy = 0,$$

and when this is subtracted from the given equation, there remains

$$R\,dx^2 + S\,dx\,dy + T\,dy^2 = dP\,dx + dQ\,dy.$$

Since $dy = -P\,dx/Q$, the differentials can be completely eliminated to indicate the relationship between x and y.

305. Let us suppose that in the solution of some problem we arrive at the equation

$$x^3d^2x + x^2y\,d^2y - y^2dx^2 + x^2dy^2 + a^2dx^2 = 0$$

and that there is no assumption about a constant differential. Since it is clear that the equation is not absurd, it follows that $x^3dx + x^2y\,dy = 0$ or $x\,dx + y\,dy = 0$, whose differential is

$$x^3d^2x + x^2y\,d^2y + 3x^2dx^2 + 2xy\,dx\,dy + x^2dy^2 = 0.$$

When this equation is subtracted from the given equation, there remains

$$a^2dx^2 - y^2dx^2 - 3x^2dx^2 - 2xy\,dx\,dy = 0,$$

or

$$a^2dx - y^2dx - 3x^2dx - 2xy\,dy = 0.$$

Since $x\,dx + y\,dy = 0$, we have

$$2xy\,dy = -2x^2dx,$$

so that $a^2dx - y^2dx - x^2dx = 0$, or $y^2 + x^2 = a^2$. Now this equation expresses a true relationship between x and y, and it agrees with the differential $x\,dx + y\,dy = 0$, which we found before. This agreement follows unless it were manifestly clear that the given equation were impossible. Since in this case that is not true, it is valid to find $x^2 + y^2 = a^2$ without integral calculus.

306. For the sake of an example of an impossible equation, let us consider

$$y^2d^2x - x^2d^2y + y\,dx^2 - x\,dy^2 + a\,dx\,dy = 0,$$

in which no constant differential is assumed. Then we would have $y^2dx - x^2dy = 0$. When this is differentiated, we have

$$y^2d^2x - x^2d^2y + 2y\,dx\,dy - 2x\,dx\,dy = 0.$$

This together with the proposed equation gives

$$y\,dx^2 - x\,dy^2 + a\,dx\,dy = 2y\,dx\,dy - 2x\,dx\,dy.$$

However, since $dy = y^2dx/x^2$, when the differentials are eliminated, we obtain

$$y - \frac{y^4}{x^3} + \frac{ay^2}{x^2} = \frac{2y^3}{x^2} - \frac{2y^2}{x},$$

or

$$x^3 - y^3 + axy = 2xy^2 - 2x^2y.$$

Now, whether this is consistent with the differential $y^2dx - x^2dy = 0$ becomes clear when it is differentiated; that is,

$$3x^2dx - 3y^2dy + ax\,dy + ay\,dx = 2y^2dx + 4xy\,dy - 2x^2dy - 4xy\,dx,$$

or

$$\frac{dy}{dx} = \frac{3x^2 + ay - 2y^2 + 4xy}{3y^2 - ax + 4xy - 2x^2}.$$

But since

$$\frac{dy}{dx} = \frac{y^2}{x^2},$$

we have

$$3x^4 + 4x^3y + ax^2y = 3y^4 + 4xy^3 - axy^2,$$

or

$$axy = \frac{3y^4 + 4xy^3 - 4x^3y - 3x^4}{x+y} = 3y^3 + xy^2 - x^2y - 3x^3.$$

From the finite equation already obtained, we have

$$axy = y^3 + 2xy^2 - 2x^2y - x^3,$$

and when this is subtracted from the previous equation there remains

$$0 = 2y^3 - xy^2 + x^2y - 2x^3,$$

which factors into $0 = y - x$ and $2y^2 + xy + 2x^2 = 0$. Of these, the equation $y = x$ can be consistent with $dy = y^2dx/x^2$, but it does not satisfy the finite equation previously found. Unless we let $a = 0$, or unless both variables x and y are set constant so that $dx = 0$ and $dy = 0$ and all of the differential equations are satisfied, the given equation cannot hold.

307. Now let us consider differential equations involving three variables x, y, and z that are of the first, second, or higher order. In order to investigate the nature of these we ought to note that a finite equation composed of three variables determines a relationship that one of them has to the other two. Hence, there is defined the kind of function that z is of x and y. A finite equation of this kind can be a solution insofar as it is clear what kind of a function of x and y is to be substituted for z in order to satisfy the equation. Likewise, a differential equation involving three variables determines what kind of function one of the variables is of the others. Hence an equation of this kind should be thought of as having been solved when the function of two variables x and y is given that when substituted for z satisfies the equation or renders it an identity. Thus a differential equation is solved if either a function z of x and y is defined or a finite equation is given by means of which the value of this same z is expressed.

308. Although every differential equation containing only two variables always expresses a determined relationship between them, nevertheless this is not always the case in differential equations in three variables. There exist equations of this kind in which it is clear that there is no possibility that some function of x and y can be substituted for z to satisfy the equation. Indeed, if the given equation is

$$z\,dy = y\,dx,$$

it is easily seen that absolutely no function of x and y can be given that when substituted for z makes $z\,dy = y\,dx$. The differentials dx and dy cannot simply vanish. In a similar way it is clear that there is no function of x and z that when substituted for y will satisfy that same equation. No matter what function of x and z might be devised for y, its differential dy contains dz, but this cannot be eliminated, since it is not in the equation. For these reasons there is no finite equation in x, y, and z that satisfies the differential equation $z\,dy = y\,dx$.

309. Hence we must distinguish among differential equations in three variables those that are imaginary and those that are real. An equation of this kind will be imaginary or absurd if there is no finite equation that could satisfy it. Such an equation is $z\,dy = y\,dx$, which we have just considered. An equation will be real if an equivalent finite equation can be exhibited in which one variable is equal to a certain function of the other two. The following equation is such an example:

$$z\,dy + y\,dz = x\,dz + z\,dx + x\,dy + y\,dx.$$

This fits with the finite equation $yz = xz + xy$, so that

$$z = \frac{xy}{y - x}.$$

We must very carefully separate these kinds of equations into imaginary and real, especially in integral calculus, since it would be ridiculous to seek an integral for a differential equation, that is, a finite equation that it would satisfy, when it is clear that none exists.

310. In the first place, then, it is clear that every differential equation with three variables in which only two differentials occur is imaginary and absurd. Let us consider an equation that contains the variable z but only the differentials dx and dy, the differential dz being completely absent. It is obvious that no function of x and y can be exhibited that can be substituted for z to produce an identical equation. Indeed, the differentials dx and dy can in no way be removed. In these cases there is absolutely no satisfactory finite equation, unless perhaps it is possible to assign a relationship between x and y that persists no matter what z might be. For example, in the equation

$$z\,dy - z\,dx = y\,dy - x\,dx,$$

which is satisfied by $y = x$. It is easy to investigate the cases in which this happens by looking for a relationship between x and y, first when $z = 0$, and then whether the same relationship persists when z has an arbitrary value.

311. Nor is it the case that an equation is absurd only if it involves three variables and two differentials. It can be absurd even if all three differentials are present. In order to show this let P and Q be functions of only x and y and consider the equation

$$dz = P\,dx + Q\,dy.$$

If this equation is not to be absurd, then z is some function of x and y whose differential is $dz = p\,dx + q\,dy$ so that $P = p$ and $Q = q$. However, we have demonstrated (paragraph 232) that $p\,dx + q\,dy$ cannot be the differential of any function of x and y unless

$$\frac{\partial p}{\partial y} = \frac{\partial q}{\partial x}.$$

Here the notation means, as we previously assumed, that $\partial p/\partial y$ is the differential of p with only y variable, divided by dy, and $\partial q/\partial x$ is the differential of q, with only x variable, divided by dx. Hence, the equation $dz = P\,dx + Q\,dy$ cannot be real unless

$$\frac{\partial P}{\partial y} = \frac{\partial Q}{\partial x}.$$

312. Absolutely the same reasoning applies to the equation

$$dZ = P\,dx + Q\,dy,$$

if Z denotes any function of z, while P and Q are functions of x and y without involving z. In order that Z might be a function of x and y, it is necessary that

$$\frac{\partial P}{\partial y} = \frac{\partial Q}{\partial x}.$$

According to this criterion, any proposed differential equation given in this general form can be judged to be either real or absurd. Hence, it is clear that the equation $z\,dz = y\,dx + x\,dy$ is real. Since $P = y$ and $Q = x$, we have

$$\frac{\partial P}{\partial y} = 1 = \frac{\partial Q}{\partial x} = 1.$$

However, the equation $az\,dz = y^2dx + x^2dy$ is absurd, since

$$\frac{\partial P}{\partial y} = 2y \qquad \text{and} \qquad \frac{\partial Q}{\partial x} = 2x.$$

But these are not equal.

313. In order to investigate this criterion more completely, let P, Q, and R be functions of x, y, and z. Every differential equation in three variables, provided that it is of the first order, is of the form

$$P\,dx + Q\,dy + R\,dz = 0.$$

Whenever this equation is real, z will be equal to some function of x and y. Furthermore, its differential will have the form $dz = p\,dx + q\,dy$. Hence, if in the given equation this function of x and y is substituted for z and if $p\,dx + q\,dy$ is substituted for dz, then of necessity, the result will be an identical equation $0 = 0$. Since from the given equation we have

$$dz = -\frac{P\,dx}{R} - \frac{Q\,dy}{R},$$

if in P, Q, and R, this function is substituted for z, then necessarily we have

$$p = -\frac{P}{R} \qquad \text{and} \qquad q = -\frac{Q}{R}.$$

314. Since $dz = p\,dx + q\,dy$, from a previous demonstration we have

$$\frac{\partial p}{\partial y} = \frac{\partial q}{\partial x}.$$

Hence, when the function in x and y is substituted for z, we have $p = -P/R$ and $q = -Q/R$, so that

$$\frac{\partial p}{\partial y} = \frac{-R\partial P + P\partial R}{R^2\partial y} \qquad \text{and} \qquad \frac{\partial q}{\partial x} = \frac{-R\partial Q + Q\partial R}{R^2\partial x}.$$

It follows that when we multiply by R^2, we obtain

$$P\frac{\partial R}{\partial y}-R\frac{\partial P}{\partial y}=Q\frac{\partial R}{\partial x}-R\frac{\partial Q}{\partial x},$$

where the denominators ∂y and ∂x indicate that in the differentials of the numerators, only that quantity that appears in the denominator is assumed to be variable. However, these differentials ∂P, ∂Q, and ∂R cannot be known until the proper value is substituted for z; since this is not known, we proceed in the following way.

315. Since P, Q, and R are functions of x, y, and z, we let

$$dP=\alpha dx+\beta dy+\gamma dz,$$
$$dQ=\delta dx+\epsilon dy+\zeta dz,$$
$$dR=\eta dx+\theta dy+\iota dz,$$

where α, β, γ, ϵ, and so forth, denote those functions that arise from differentiation. Now let us consider the substitution everywhere for z the function in x and y that is equal to z, and for dz we substitute the expression $p\,dx+q\,dy$ and thus obtain

$$dP=(\alpha+\gamma p)\,dx+(\beta+\gamma q)\,dy,$$
$$dQ=(\delta+\zeta p)\,dx+(\epsilon+\zeta q)\,dy,$$
$$dR=(\eta+\iota p)\,dx+(\theta+\iota q)\,dy.$$

From these equations we obtain

$$\frac{\partial R}{\partial y}=\theta+\iota q,\qquad \frac{\partial R}{\partial x}=\eta+\iota p,$$

$$\frac{\partial P}{\partial y}=\beta+\gamma q,\qquad \frac{\partial Q}{\partial x}=\delta+\zeta p.$$

316. Since the reality of the equation requires that

$$P\frac{\partial R}{\partial y}-R\frac{\partial P}{\partial y}=Q\frac{\partial R}{\partial x}-R\frac{\partial Q}{\partial x},$$

the result is that when we substitute the values just found, we obtain

$$P\,(\theta+\iota q)-R\,(\beta+\gamma q)=Q\,(\eta+\iota p)-R\,(\delta+\zeta p)\,.$$

However, we have already seen that $p=-P/R$ and $q=-Q/R$. But these values can be obtained, even if the function in x and y is not substituted for

z, since the differentials are no longer required in the computation. Hence we have

$$P\theta - \frac{PQ\iota}{R} - R\beta + Q\gamma = Q\eta - \frac{PQ\iota}{R} - R\delta + P\zeta,$$

or

$$0 = P(\zeta - \theta) + Q(\eta - \gamma) + R(\beta - \delta).$$

Since the quantities β, δ, γ, η, ζ, θ, were found by differentiation, when we use the notation given previously, we have

$$0 = P\left(\frac{\partial Q}{\partial z} - \frac{\partial R}{\partial y}\right) + Q\left(\frac{\partial R}{\partial x} - \frac{\partial P}{\partial z}\right) + R\left(\frac{\partial P}{\partial y} - \frac{\partial Q}{\partial x}\right).$$

Unless this condition is met, the original equation is not real, but imaginary and absurd.

317. Although we have discovered this rule from a consideration of the variable z, still, since all of the variables enter in equally, it is clear that from a consideration of the other variables, the same expression will result. Hence, if a first-order differential equation involving three variables is proposed, we can determine immediately whether it is real or imaginary. Indeed, it can be put into the general form

$$P\,dx + Q\,dy + R\,dz$$

and then we investigate the value of the formula

$$P\left(\frac{\partial Q}{\partial z} - \frac{\partial R}{\partial y}\right) + Q\left(\frac{\partial R}{\partial x} - \frac{\partial P}{\partial z}\right) + R\left(\frac{\partial P}{\partial y} - \frac{\partial Q}{\partial x}\right).$$

If this is equal to zero, then the equation is real; if it is not equal to zero, then the equation is imaginary or absurd.

318. A given equation can always be reduced by division to the form

$$P\,dx + Q\,dy + dz = 0.$$

If $R = 1$, the previous criterion becomes simpler:

$$P\frac{\partial Q}{\partial z} - Q\frac{\partial P}{\partial z} + \frac{\partial P}{\partial y} - \frac{\partial Q}{\partial x} = 0.$$

Whenever this expression is really equal to zero, the given equation is real; if the contrary is true, then the equation is imaginary. After all of this has been demonstrated, it is certain. However, a priori, it could be doubted whether an equation is always real whenever this criterion so indicated. Since at this time we cannot demonstrate this completely, nevertheless in integral calculus this can be confirmed. At this time we simply affirm that this is true and that no one should fear any danger from this, even if in the meantime someone wishes to entertain some doubt about its truth.

319. From this criterion, in the first place, it is clear that if in the equation

$$P\,dx + Q\,dy + R\,dz = 0$$

P is a function only of x, Q is a function only of y, and R is a function only of z, then the equation will always be real. Indeed, since

$$\frac{\partial P}{\partial y} = 0, \quad \frac{\partial P}{\partial z} = 0, \quad \frac{\partial Q}{\partial z} = 0, \quad \frac{\partial Q}{\partial x} = 0, \quad \frac{\partial R}{\partial x} = 0, \quad \frac{\partial R}{\partial y} = 0,$$

the whole expression vanishes spontaneously.

320. If, as before, P is a function of x and Q is a function of y, but only R is a function of x, y, and z, then the equation will we real if

$$P\frac{\partial R}{\partial y} = Q\frac{\partial R}{\partial x}, \qquad \text{or} \qquad \frac{\partial R}{\partial x}\Big/\frac{\partial R}{\partial y} = \frac{P}{Q}.$$

For example, if the given equation is

$$\frac{2dx}{x} + \frac{3dy}{y} + \frac{x^2y^3dz}{z^6} = 0,$$

since here

$$P = \frac{2}{x}, \qquad Q = \frac{3}{y}, \qquad R = \frac{x^2y^3}{z^6},$$

we have

$$\frac{\partial R}{\partial x} = \frac{2xy^3}{z^6} \qquad \text{and} \qquad \frac{\partial R}{\partial y} = \frac{3x^2y^2}{z^6},$$

and so

$$P\frac{\partial R}{\partial y} = Q\frac{\partial R}{\partial x} = \frac{6xy^2}{z^6}.$$

It follows that the given equation is real.

321. If P and Q are functions of x and y, while R is a function of z alone, since

$$\frac{\partial P}{\partial z} = 0, \qquad \frac{\partial Q}{\partial z} = 0, \qquad \frac{\partial R}{\partial x} = 0, \qquad \frac{\partial R}{\partial y} = 0,$$

the equation will be real, provided that

$$\frac{\partial P}{\partial y} = \frac{\partial Q}{\partial x}.$$

This same condition is required if $P\,dx+Q\,dy$ is to be a determined differential, that is, one that arises from the differentiation of some finite function of x and y. This brings us back to what we have previously observed in paragraph 312, that the equation $dZ = P\,dx + Q\,dy$ with Z a function of z alone, while P and Q are functions of x and y, can be real only if

$$\frac{\partial P}{\partial y} = \frac{\partial Q}{\partial x}.$$

Both of these cases are interrelated, since if R is a function of z alone we can substitute dZ for $R\,dz$, where Z is a function of z.

322. In order to illustrate this criterion let us consider the following equation:

$$\left(6xy^2z - 5yz^3\right) dx + \left(5x^2yz - 4xz^3\right) dy + \left(4x^2y^2 - 6xyz^2\right) dz = 0.$$

When this is compared to the general form, we obtain

$$P = 6xy^2z - 5yz^3, \qquad \frac{\partial P}{\partial y} = 12xyz - 5z^3, \qquad \frac{\partial P}{\partial z} = 6xy^2 - 15yz^2,$$

$$Q = 5x^2yz - 4xz^3, \qquad \frac{\partial Q}{\partial x} = 10xyz - 4z^3, \qquad \frac{\partial Q}{\partial z} = 5x^2y - 12xz^2,$$

$$R = 4x^2y^2 - 6xyz^2, \qquad \frac{\partial R}{\partial x} = 8xy^2 - 6yz^2, \qquad \frac{\partial R}{\partial y} = 8x^2y - 6xz^2.$$

With these values discovered, the equation giving the solution is

$$\left(6xy^2z - 5yz^3\right)\left(-3x^2y - 6xz^2\right) + \left(5x^2yz - 4xz^3\right)\left(2xy^2 + 9yz^2\right)$$
$$+ \left(4x^2y^2 - 6xyz^2\right)\left(2xyz - z^3\right) = 0.$$

But when this expression is simplified, each term is negated by another, so that $0 = 0$, which indicates that the given equation is real.

323. When the expression obtained in this way from the criterion fails to vanish, this is a sign that the given equation is imaginary. However, when a finite equation is found in this way from the criterion, provided that it is consistent with the differential equation, it indicates the relationship that the variables have to each other. Furthermore, this is the way in which those cases arise that we recall from paragraph 310. Suppose that the given equation is

$$(z - x)\,dx + (y - z)\,dy = 0.$$

Then

$$P = z - x, \qquad Q = y - z, \qquad \text{and} \qquad R = 0,$$

but

$$\frac{\partial P}{\partial z} = 1 \qquad \text{and} \qquad \frac{\partial Q}{\partial z} = -1.$$

The deciding equation becomes

$$P\frac{\partial Q}{\partial z} = Q\frac{\partial P}{\partial z},$$

or

$$z - x = z - y,$$

so that

$$y = x.$$

Since in this case it turns out that $y = x$ also satisfies the differential equation, we have to say that the given differential equation has no other significance than $y = x$.

324. Hence, when a differential equation containing three variables is given,

$$P\,dx + Q\,dy + R\,dz = 0,$$

there are the three following cases that must be considered concerning the equation which results:

$$P\left(\frac{\partial Q}{\partial z} - \frac{\partial R}{\partial y}\right) + Q\left(\frac{\partial R}{\partial x} - \frac{\partial P}{\partial z}\right) + R\left(\frac{\partial P}{\partial y} - \frac{\partial Q}{\partial x}\right) = 0.$$

The first case occurs if this expression is really equal to zero, and then the given equation is real. However, if this finite equation is not an identity, then it must be decided whether it satisfies the given equation. If this happens, we have a finite equation, and this is the second case. The third case occurs if the finite equation does not agree with the given differential equation, and then the given equation is imaginary. In this case no finite equation can be found that satisfies the given equation.

325. The first and third cases are self-evident. The second, however, although quite rare, deserves special consideration. Since the example already considered contains only two differentials, we will give another equation, which has all three differentials:

$$(z - y)\,dx + x\,dy + (y - z)\,dz = 0.$$

Here we have

$$P = z - y, \qquad \frac{\partial Q}{\partial z} = 0, \qquad \frac{\partial R}{\partial y} = 1,$$

$$Q = x, \qquad \frac{\partial R}{\partial x} = 0, \qquad \frac{\partial P}{\partial z} = 1,$$

$$R = y - z, \qquad \frac{\partial P}{\partial y} = -1, \qquad \frac{\partial Q}{\partial x} = 1,$$

so that the finite equation resulting from the criterion is $z - x - y = 0$, or $z = x + y$. When this value for z is substituted in the differential equation we have

$$x\,dx + x\,dy - x\,(dx + dy) = 0.$$

Since this equation is an identity, it follows that the differential equation signifies nothing but $z = x + y$.

326. Since we have said that all first-order differential equations containing three variables are of the form

$$P\,dx + Q\,dy + R\,dz = 0,$$

some question may arise concerning those equations in which the first differentials are raised to the second or higher power. For example,

$$P\,dx^2 + Q\,dy^2 + R\,dz^2 = 2S\,dx\,dy + 2T\,dx\,dz + 2V\,dy\,dz.$$

It should be noted about equations of this kind that they could not possibly be real unless they have divisors of the previous form that make up simple equations. Since from the given equation we have

$$dz = \frac{T\,dx + V\,dy}{R}$$
$$\pm \frac{\sqrt{dx^2\,(T^2 - PR) + 2dx\,dy\,(TV + RS) + dy^2\,(V^2 - QR)}}{R},$$

it is clear that z cannot be a function or x and y, nor does dz have the form $p\,dx + q\,dy$ unless the irrational expression turns out to be rational. This happens if

$$\left(T^2 - PR\right)\left(V^2 - QR\right) = (TV + RS)^2,$$

or

$$R = \frac{PV^2 + 2STV + QT^2}{PQ - S^2}.$$

Hence unless this finite equation is satisfied, the given equation will be imaginary.

327. We might have treated in this chapter differential equations of higher order that contain three variables, and we might have considered and decided which of these turn out to be either real or imaginary. However, since the criteria become extremely intricate, we omit this work, especially since this follows from the same sources which we have here explored. Indeed, if there is need for these criteria in integral calculus, at that stage they can easily be developed. For the same reason we have not at this time considered equations with more variables, especially since they practically never occur. If it is ever necessary, there should be no difficulty in examining such equations with the principles we have discussed here. For these reasons we here bring to a conclusion our exposition of the principles of differential calculus. We next move on to show some of the more important applications that this calculus has both in analysis and in higher geometry.

Index